Mechanics of Aeronautical Solids, Materials and Structures

Mechanics of Aeronautical Solids, Materials and Structures

Christophe Bouvet

iSTE

WILEY

First published 2017 in Great Britain and the United States by ISTE Ltd and John Wiley & Sons, Inc.

ISTE Ltd
27-37 St George's Road
London SW19 4EU
UK

www.iste.co.uk

John Wiley & Sons, Inc.
111 River Street
Hoboken, NJ 07030
USA

www.wiley.com

Library of Congress Control Number: 2016962010

British Library Cataloguing-in-Publication Data
A CIP record for this book is available from the British Library
ISBN 978-1-78630-115-4

Contents

Foreword

This book follows a long-standing tradition of mechanical engineering tuition, which is already a century old and comes from the Toulouse mechanical engineering scientific community, now merged into the Institut Clément Ader (www.institut-clement-ader.org). Just as all of its illustrious predecessors, this book is very timely and illustrates the specificity and originality of the approaches that we have developed, which have both a high scientific standing and a quasi-permanent connection with the aeronautical industry. This publication provides the reader with the necessary knowledge and techniques to calculate structures and decompartmentalize disciplines and fields. The aeronautical engineer will find all of the helpful information he or she needs within these pages: the basis of continuum mechanics, the finite element method, and knowledge of materials, metals and composites, both within linear and non-linear fields. The information is presented in an extremely clear and educational manner. The reader may draw on an impressive series of exercises with detailed corrections, something which is not so commonly found.

Bruno CASTANIÉ
INSA Toulouse
Institut Clément Adler

Preface

This volume, on the mechanics of solids and materials, as well as aeronautical structures, aims to give an overview of the necessary notions for structure sizing within the aeronautics field. It begins by establishing all of the classic notions of mechanics: stress, strain, behavior law and sizing criteria. Also covered are notions that are specific to aeronautics, with a particular emphasis on the notion of limit loads and ultimate loads.

Different problem-solving methods, particularly the finite element method, are then introduced. The methods are not classically presented and instead energy minimization is drawn on in order to minimize the number of equations, all while remaining within a framework that we may comprehend "with their hands".

The book then addresses the subject of plasticity, showcasing its influence on structure sizing, and especially the advantages it has for sizing criteria.

Finally, the physics of the two main materials in aeronautics, namely aluminum and composite materials, is discussed, so as to shed light on the sizing criteria outlined in the previous chapters.

The corrected exercises help the student to test their understanding of the different topics.

What is so original about this book is that from the outset, it places itself within the field of aeronautics. Sizing criteria are indeed rather specific to this field. Nevertheless, the notions discussed remain valid for the majority of industrial fields: in Mechanical Engineering and Finite Elements these notions in fact remain the same.

Another original aspect of this work is that it consolidates basic continuum mechanics with a very succinct description of finite elements, and a description of the material aspect of the main materials used in aeronautical structures, that being aluminum and composites. This publication is therefore a summary of the basic knowledge deemed necessary for the ("Airbus") engineer working within research departments. The book is simultaneously aimed at both students who are beginning their training and also engineers already working in the field who desire a summary of the basic theories.

Lastly, the publication aims to limit the amount of formulas provided as much as possible, in order to highlight the significance of the physical. Any readers who may be interested in demonstrations are advised to refer to more specific and theoretical works, such as [COI 01, DUV 98, GER 73, HEA 77, KHA 95, LEM 96, MIR 03, SAL 01, UGU 03] and [THU 97], etc.

Christophe BOUVET
January 2017

Introduction

I.1. Outlining the problem

Let us consider a solid S that is subjected to imposed displacements and external forces.

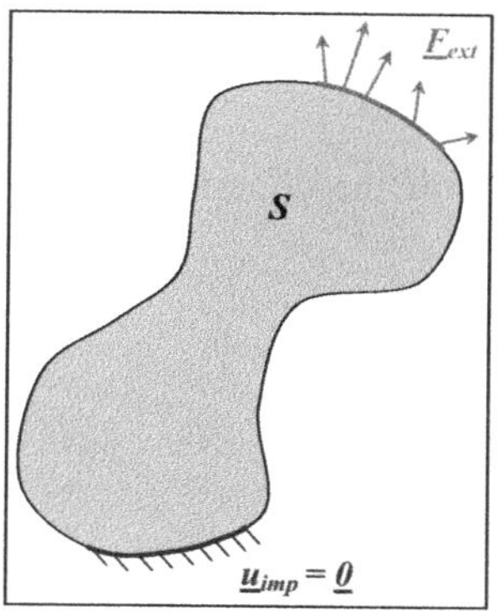

Figure I.1. *Outlining the problem. For a color version of this figure, see www.iste.co.uk/bouvet/aeronautical.zip*

The aim of the mechanics of deformable solids is to study the internal state of the material (notion of stress) and the way in which it becomes deformed (notion of strain) [FRA 12, SAL 01, LEM 96].

In mechanics, a mechanical piece or system may be designed:

– to prevent it from breaking;

– to prevent it from becoming permanently deformed;

– to prevent it from becoming too deformed, or;

– for any another purposes.

A solid shall be deemed a continuous medium, meaning that it shall be regarded as a continuous set of material points with a mass, representing the state of matter that is surrounded by an infinitesimal volume.

Mechanics of deformable solids enables the study of cohesive forces (notion of stress) at a point M, like the forces exerted on the small volume surrounding it, called a Representative Elementary Volume (REV). For metals, the REV is typically within the range of a tenth of a millimeter.

The matter in this REV must be seen as continuous and homogeneous:

– if it is too small, the matter cannot be as seen homogeneous: atomic piling, inclusion within matter, grains, etc. (for example: for concrete, the REV is within the range of 10 cm);

– if it is too big, the state of the cohesive forces in its center will no longer represent the REV state.

1

Stress

1.1. Notion of stress

1.1.1. *External forces*

There are three types of external forces:

– concentrated forces: this is a force exerted on a point (in Newton units, noted as N). In practice, this force does not actually exist. It is just a model. If we were to apply a force to a point that has zero surface, the contact pressure would be infinite and the deformation of the solid would therefore induce a non-zero contact surface. Nevertheless, it can still be imagined for studying problems with a very concentrated contact type load between balls. The results will thus yield an infinite stress and will need to be interpreted accordingly;

– surface forces, which will be noted as $\underline{F}_{ext}$ for the rest of this volume (in Pascal units, it is noted Pa). This type of force includes contact forces between two solids as well as the pressure of a fluid. Practically, any concentrated force can be seen as a surface force distributed onto a small contact surface;

– volume forces, which will be noted as $\underline{f}_v$ for the rest of this book (in N/m^3). Examples of volume forces are forces of gravity, electromagnetic forces, etc.

Incidentally, in this book you will notice that vectors are underlined once and matrices (or tensors of rank 2), which you will come across further on, are underlined twice.

1.1.2. *Internal cohesive forces*

We wish to study the cohesive forces of the solid S, at point M and which is in equilibrium under the action of external forces. The solid is cut into two parts E_1 and E_2 by a plane with a normal vector $\underline{n}$ passing through M. The part E_1 is in equilibrium under the action of the external forces on E_1 and the cohesive force of E_2 on E_1.

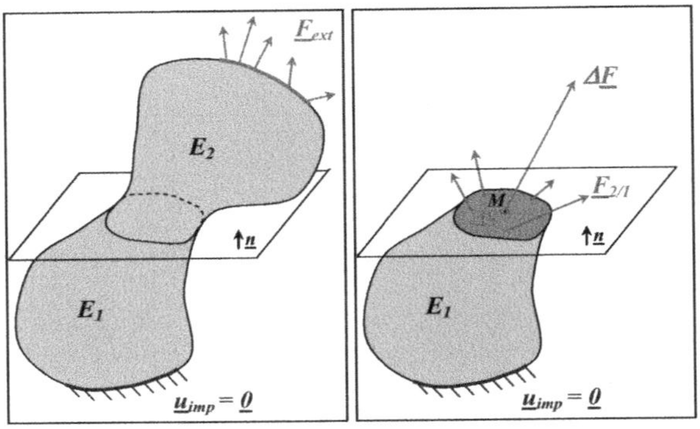

Figure 1.1. *Principle of internal cohesive forces. For a color version of this figure, see www.iste.co.uk/bouvet/aeronautical.zip*

Let ΔS be the surface around M and $\underline{\Delta F}$ be the cohesive force of 2 on 1 exerting on ΔS, then the stress vector at the point M associated with the facet with a normal vector $\underline{n}$ is called:

$$\underline{\sigma}(M,\underline{n}) = \lim_{\Delta S \to 0} \frac{\Delta \underline{F}}{\Delta S} = \frac{d\underline{F}}{dS} \qquad [1.1]$$

The unit is N/m^2 or Pa and we generally use MPa or N/mm^2.

Physically, the stress notion is fairly close to the notion of pressure that can be found in everyday life (the unit is even the same!), but as we will see further on, pressure is but only one particular example of stress.

1.1.3. *Normal stress, shear stress*

We define the different stresses as:

– normal stress, the projection of $\underline{\sigma}(M, \underline{n})$ onto $\underline{n}$, noted as σ;

– shear stress, the projection of $\underline{\sigma}(M, \underline{n})$ onto the plane with normal n, noted as τ.

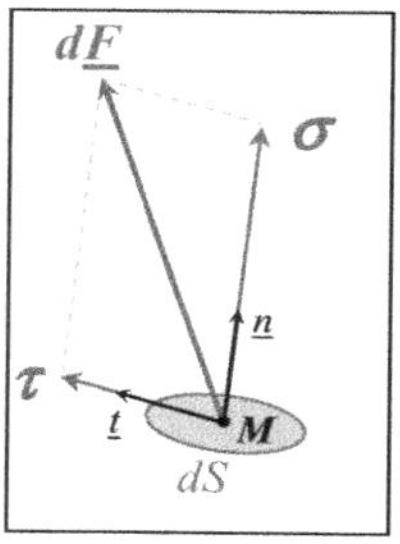

Figure 1.2. *Decomposition of a stress vector. For a color version of this figure, see www.iste.co.uk/bouvet/aeronautical.zip*

Thus, σ represents the cohesive forces perpendicular to the facet, meaning the traction/compression, and τ the forces tangential to the facet, meaning the shear. In a physical sense, the pressure found in our everyday lives is simply a normal compression stress.

We then definitely have:

$$\underline{\sigma}(M,\underline{n}) = \sigma\underline{n} + \tau\underline{t} \tag{1.2}$$

NOTE.– *$\underline{n}$ and $\underline{t}$ must be unit vectors.*

And conversely:

$$\begin{cases} \sigma = \underline{\sigma}(M,\underline{n}) \cdot \underline{n} \\ \tau = \underline{\sigma}(M,\underline{n}) \cdot \underline{t} \end{cases} \tag{1.3}$$

1.2. Properties of the stress vector

1.2.1. *Boundary conditions*

If $\underline{n}$ is an external normal, then:

$$\underline{\sigma}(M,\underline{n}_{ext}) = \underline{F}_{ext} \tag{1.4}$$

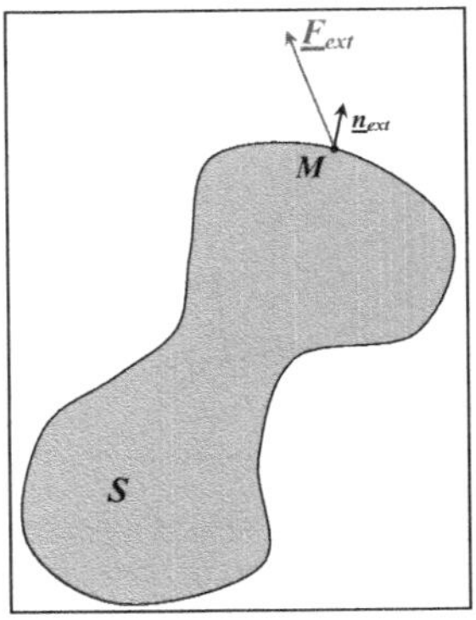

Figure 1.3. *External force and associated normal vector. For a color version of this figure, see www.iste.co.uk/bouvet/aeronautical.zip*

NOTE.– *$\underline{F}_{ext}$ is in MPa, and a normal external vector is always moving from the matter towards the exterior.*

So, $\underline{F}_{ext}$ can be seen as a stress vector exerted on S, particularly if the surface is a free surface:

$$\underline{\sigma}(M, \underline{n}_{ext}) = \underline{0} \qquad [1.5]$$

These relations are important as they translate the stress boundary conditions on the structure. In order for this to be the solution to the problem (see Chapter 3), these relations are part of a group of conditions that are needed to verify a stress field.

EXAMPLE: TANK UNDER PRESSURE.–

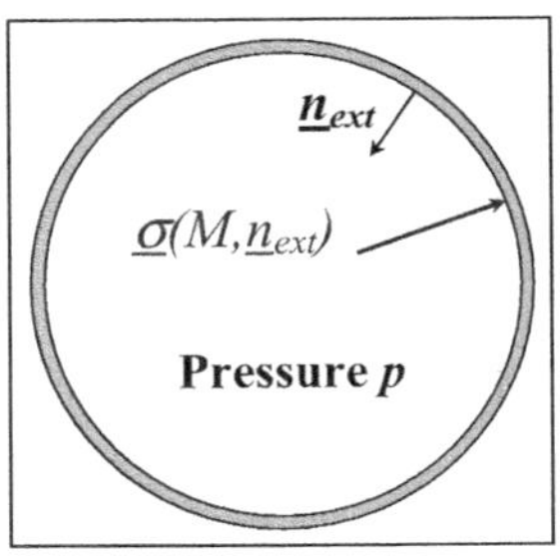

Figure 1.4. *Tank under pressure*

For every point on the internal wall of the tank, we find:

$$\underline{\sigma}(M,\underline{n}_{ext}) = -p.\underline{n}_{ext} \qquad [1.6]$$

With the external normal vector moving towards the center of the circle, from where the normal and shear stresses are:

$$\begin{cases} \sigma = \underline{\sigma}(M,\underline{n}_{ext}).\underline{n}_{ext} = -p \\ \tau = \underline{\sigma}(M,\underline{n}_{ext}).\underline{t} = 0 \end{cases} \qquad [1.7]$$

Given that the normal stress is negative and the shear stress is zero, the material is subjected to pure compression. The first relation shows that the physical notion of pressure is simply a normal stress of compression: hence the minus sign before the pressure!

1.2.2. *Torsor of internal forces*

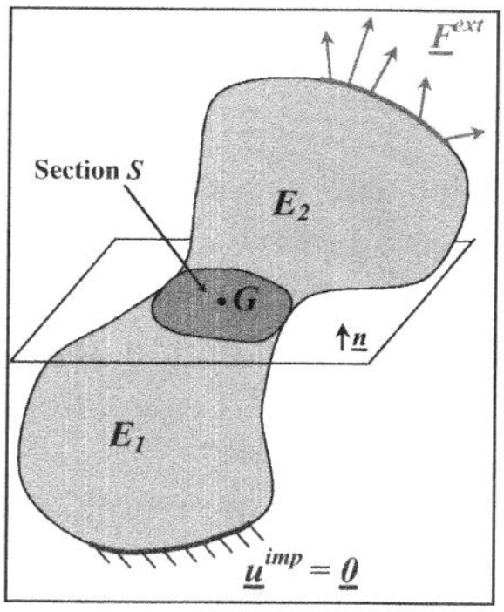

Figure 1.5. *Set of internal forces. For a color version of this figure, see www.iste.co.uk/bouvet/aeronautical.zip*

The torsor of internal forces of 2 on 1 at G, the center of gravity of S, is:

$$\left\{ T_{2/1}^{coh} \right\}_G = \left\{ \begin{matrix} \underline{R}_{2/1} \\ \underline{M}_{2/1}(G) \end{matrix} \right\} \qquad [1.8]$$

At first sight, the torsor notion may seem primitive but it enables us to simply consolidate the force with the moment. Should the notion of torsor bother you, you may settle for referring to it in plainer language as force and moment. However, you should not forget that when speaking about internal forces between 2 parts of a solid, it needs to be remembered that there is a force (in N) and a moment (in N.mm). The ambiguity comes from the term "force", which is used for a force (in the common everyday sense of the word), and as a whole, force + moment!

Let us now seek to link this set of internal forces to the previously discussed stress vector. We then have:

$$d\underline{F}_{2/1}(M) = \underline{\sigma}(M,\underline{n}).dS \qquad [1.9]$$

therefore:

$$\begin{cases} \underline{R}_{2/1} = \sum\limits_{M \in S} d\underline{F}_{2/1}(M) = \iint_S \underline{\sigma}(M,\underline{n}).dS \\ \underline{M}_{2/1}(G) = \sum\limits_{M \in S} d\underline{M}_{2/1}^{(M)}(G) = \sum\limits_{M \in S} \underline{GM} \wedge d\underline{F}_{2/1}(M) \\ \qquad\qquad = \iint_S \underline{GM} \wedge \underline{\sigma}(M,\underline{n}).dS \end{cases} \qquad [1.10]$$

These relations are somewhat (or very) complex, but physically, they simply translate the fact that if we add up all of the stress vectors on section S, then we will obtain the force of part E_2 on part E_1. Lastly, we should not forget that when we add up the stress vectors, we will obtain not only a force, but also a moment (which obviously depends on the point at which it is calculated).

These relations can also be written on an external surface as:

$$\begin{cases} \underline{R}_{ext/1} = \iint_{S_{ext}} \underline{\sigma}(M,\underline{n}_{ext}).dS = \iint_{S_{ext}} \underline{F}_{ext}.dS \\ \underline{M}_{ext/1}(G) = \iint_{S_{ext}} \underline{GM} \wedge \underline{\sigma}(M,\underline{n}_{ext}).dS = \iint_{S_{ext}} \underline{GM} \wedge \underline{F}_{ext}.dS \end{cases} \qquad [1.11]$$

These relations are important because in practice, although we know the resultant $\underline{R}_{ext/1}$ or $\underline{M}_{ext/1}$, we do not generally know $\underline{F}_{ext}$. In fact, an external force is practically applied via the intermediary of a beam, a screed, a jack, etc., and the applied resulting force (or the moment) is known, but the way in which it is divided is unknown.

EXAMPLE: TRACTION.–

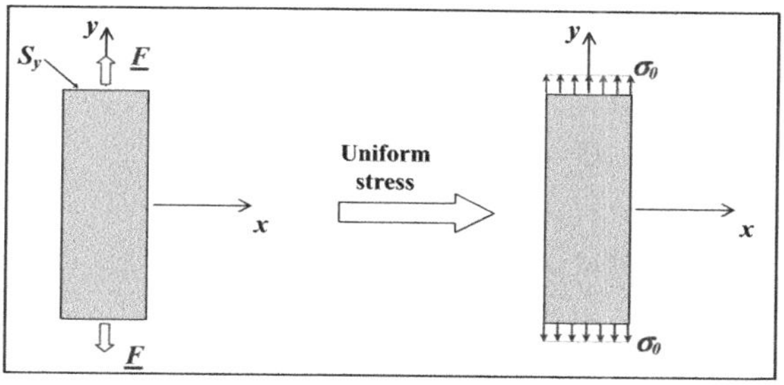

Figure 1.6. *Tensile test. For a color version of this figure, see www.iste.co.uk/bouvet/aeronautical.zip*

In a tensile test, we know that the resultant of the forces applied to S_y is worth $\underline{F}$:

$$\begin{cases} \iint_{S_y} \underline{\sigma}(M,\underline{y}).dS = \underline{F} \\ \iint_{S_y} \underline{GM} \wedge \underline{\sigma}(M,\underline{y}).dS = \underline{0} \end{cases} \qquad [1.12]$$

However, in order to deduce that:

$$\underline{\sigma}(M,\underline{y}) = \frac{F}{S_y} = \sigma_0 \cdot \underline{y} \qquad [1.13]$$

we must add a homogeneity hypothesis of the force applied which remains to be verified. Incidentally, we can demonstrate that the two previous integrals are verified with this stress vector.

EXAMPLE: BENDING.–

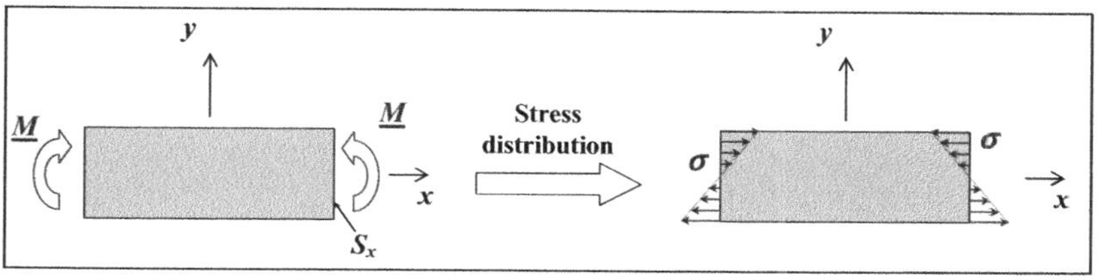

Figure 1.7. *Bending test. For a color version of this figure, see www.iste.co.uk/bouvet/aeronautical.zip*

In a pure bending test, we know that the resultant of the forces applied to S_x is worth $M.\underline{z}$:

$$\begin{cases} \iint_{S_X} \underline{\sigma}(M,\underline{x}).dS = \underline{0} \\ \iint_{S_X} \underline{GM} \wedge \underline{\sigma}(M,\underline{x}).dS = M.\underline{z} \end{cases} \quad [1.14]$$

However, by deducing that on S_x:

$$\underline{\sigma}(M,\underline{x}) = \frac{-M}{I_z}.y.\underline{x} \quad [1.15]$$

This formula is a classic example of the mechanics of material which we will discuss (and demonstrate) again when doing the exercises. Should you need to, you can read a more detailed publication, such as [AGA 08, BAM 08, CHE 08, DEL 08, DUP 09], etc.

Obviously, with the moment of inertia:

$$I_z = \iint_{S_X} y^2.dS \quad [1.16]$$

we must add a linear distribution hypothesis of the stress applied which remains to be verified. Incidentally, we can demonstrate that the two previous integrals are verified with this stress vector.

1.2.3. *Reciprocal actions*

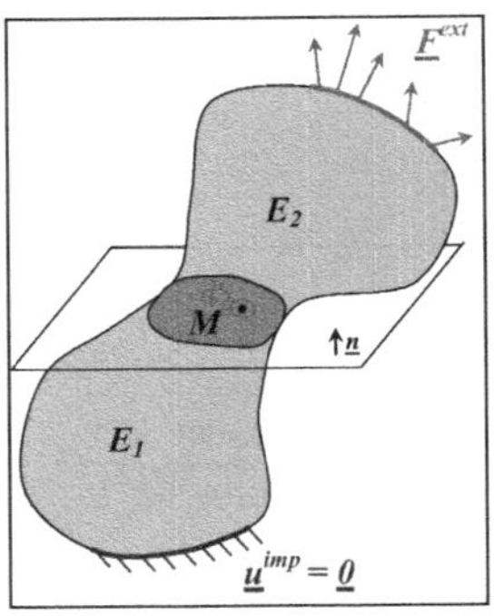

Figure 1.8. *Reciprocal actions. For a color version of this figure, see www.iste.co.uk/bouvet/aeronautical.zip*

According to the Law of Reciprocal Action, we have:

$$d\underline{F}_{2/1} = -d\underline{F}_{1/2} \qquad [1.17]$$

Yet:

$$\begin{cases} d\underline{F}_{2/1}(M) = \underline{\sigma}(M,\underline{n}).dS \\ d\underline{F}_{1/2}(M) = \underline{\sigma}(M,-\underline{n}).dS \end{cases} \qquad [1.18]$$

Hence:

$$\underline{\sigma}(M,\underline{n}) = -\underline{\sigma}(M,-\underline{n}) \qquad [1.19]$$

This can be translated by the fact that a fine slice of matter of surface dS, which has a normal vector $+\underline{n}$ on one side and $-\underline{n}$ on the other, is at equilibrium under the action of the two opposing forces $\underline{\sigma}(M, \underline{n}).dS$ and $\underline{\sigma}(M, -\underline{n}).dS$. Evidently, it is very much at equilibrium.

1.2.4. *Cauchy reciprocal theorem*

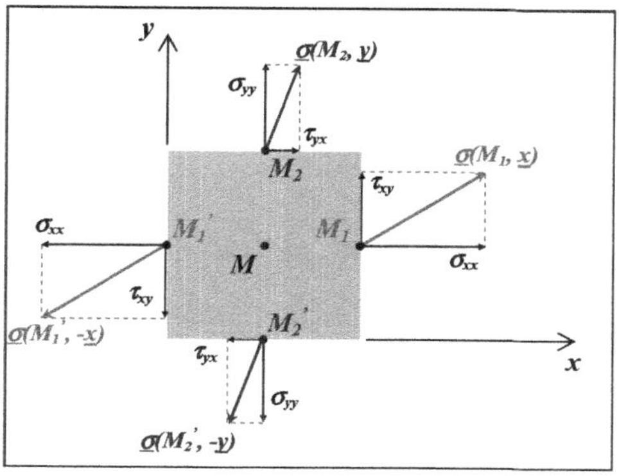

Figure 1.9. *Stress vectors on the faces of a square. For a color version of this figure, see www.iste.co.uk/bouvet/aeronautical.zip*

Let us put this in 2D, in order to make the demonstration easier.

A square is assumed to be infinitely small, therefore the stresses are assumed to be constant everywhere in the cube, hence we have:

$$\begin{cases} \underline{\sigma}\left(M_1,\underline{x}\right)=\underline{\sigma}\left(M,\underline{x}\right)=\sigma_{xx}\cdot\underline{x}+\tau_{xy}\cdot\underline{y} \\ \underline{\sigma}\left(M_2,\underline{y}\right)=\underline{\sigma}\left(M,\underline{y}\right)=\sigma_{yy}\cdot\underline{y}+\tau_{yx}\cdot\underline{x} \\ \underline{\sigma}\left(M_1',-\underline{x}\right)=-\underline{\sigma}\left(M,\underline{x}\right)=-\sigma_{xx}\cdot\underline{x}-\tau_{xy}\cdot\underline{y} \\ \underline{\sigma}\left(M_2',-\underline{y}\right)=-\underline{\sigma}\left(M,\underline{y}\right)=-\sigma_{yy}\cdot\underline{y}-\tau_{yx}\cdot\underline{x} \end{cases} \quad [1.20]$$

In the notation of τ_{xy}, the first "x" corresponds to the direction of the facet, meaning the normal vector on the cutting plane in question, and the second index "y" represents the direction of the stress.

The equilibrium equation on the square is written as:

$$\Sigma\left\{T^{coh}_{ext/cube}\right\}=\{0\} \quad [1.21]$$

which, for the force equation, induces the following:

$$\underline{\sigma}\left(M_1,\underline{x}\right).dy.dz+\underline{\sigma}\left(M_2,\underline{y}\right).dx.dz+\underline{\sigma}\left(M_1',-\underline{x}\right).dy.dz+\underline{\sigma}\left(M_1',-\underline{y}\right).dx.dz=\underline{0} \quad [1.22]$$

This is an automatically verified equation. For the moment equation in M at the center of the square, the below is induced:

$$\begin{aligned} &\underline{MM_1}\wedge\underline{\sigma}\left(M_1,\underline{x}\right).dy.dz+\underline{MM_2}\wedge\underline{\sigma}\left(M_2,\underline{y}\right).dx.dz \\ &\quad+\underline{MM_1'}\wedge\underline{\sigma}\left(M_1',-\underline{x}\right).dy.dz+\underline{MM_2'}\wedge\underline{\sigma}\left(M_2',-\underline{y}\right).dx.dz=\underline{0} \end{aligned} \quad [1.23]$$

where the Cauchy reciprocity theorem is:

$$\tau_{xy}=\tau_{yx} \quad [1.24]$$

It can be shown in the same way in 3D:

$$\begin{cases} \tau_{xy}=\tau_{yx} \\ \tau_{xz}=\tau_{zx} \\ \tau_{yz}=\tau_{zy} \end{cases} \quad [1.25]$$

EXAMPLE: TORSION OF A WELDED TUBE.–

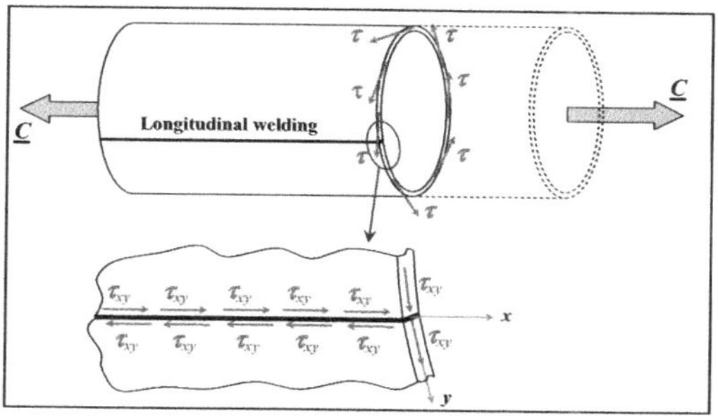

Figure 1.10. *Torsion of a welded tube. For a color version of this figure, see www.iste.co.uk/bouvet/aeronautical.zip*

A torsion moment is applied to a tube comprised of a rolled and welded sheet. It can be shown (see exercises) that it has been subjected to a homogeneous circumferential shear stress. The Cauchy reciprocity theorem then induces the welding to be stressed by a longitudinal shear τ equal to the circumferential shear stress. All that remains then is the sizing of the welding so that it can withstand this shearing applied force.

1.3. Stress matrix

1.3.1. *Notation*

There is a stress vector on the facet with the normal vector $\underline{x}$:

$$\underline{\sigma}(M,\underline{x})=\underline{\sigma}+\underline{\tau}=\sigma.\underline{x}+\tau.\underline{t} \qquad [1.26]$$

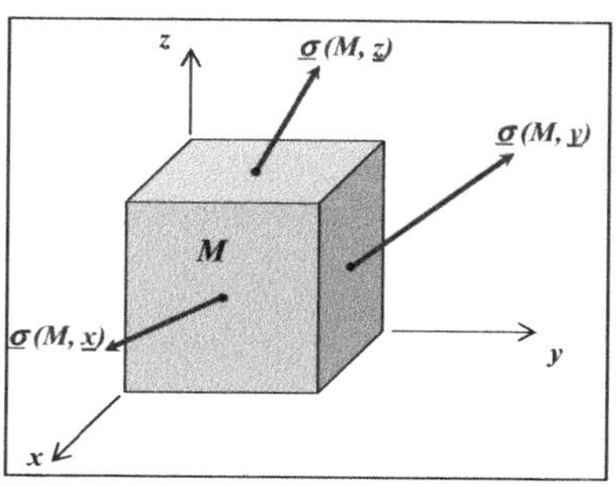

Figure 1.11. *Stress vectors on the faces of a unit cube. For a color version of this figure, see www.iste.co.uk/bouvet/aeronautical.zip*

and $\underline{t}$ can be broken down in accordance with $\underline{y}$ and $\underline{z}$, hence:

$$\underline{\sigma}(M,\underline{x}) = \sigma_{xx} \cdot \underline{x} + \tau_{xy} \cdot \underline{y} + \tau_{xz} \cdot \underline{z} \qquad [1.27]$$

Once again, in the notation of τ_{xy}, the first index "x" corresponds to the direction of the facet and the second index "y" represents the direction of the stress.

In the same way for the faces of the cube with normal vectors $\underline{y}$ and $\underline{z}$, we have:

$$\begin{cases} \underline{\sigma}(M,\underline{x}) = \sigma_{xx} \cdot \underline{x} + \tau_{xy} \cdot \underline{y} + \tau_{xz} \cdot \underline{z} \\ \underline{\sigma}(M,\underline{y}) = \tau_{yx} \cdot \underline{x} + \sigma_{yy} \cdot \underline{y} + \tau_{yz} \cdot \underline{z} \\ \underline{\sigma}(M,\underline{z}) = \tau_{zx} \cdot \underline{x} + \tau_{zy} \cdot \underline{y} + \sigma_{zz} \cdot \underline{z} \end{cases} \qquad [1.28]$$

$\underline{\underline{\sigma}}(M)$ is what is referred to as the stress matrix:

$$\underline{\underline{\sigma}}(M) = \begin{bmatrix} \sigma_{xx} & \tau_{yx} & \tau_{zx} \\ \tau_{xy} & \sigma_{yy} & \tau_{zy} \\ \tau_{xz} & \tau_{yz} & \sigma_{zz} \end{bmatrix}_{(\underline{x},\underline{y},\underline{z})} \qquad [1.29]$$

Evidently, the expression of this matrix depends on the coordinate.

Moreover, it is symmetrical in accordance with the reciprocity of the stresses:

$$\underline{\underline{\sigma}}(M) = \begin{bmatrix} \sigma_{xx} & \tau_{xy} & \tau_{xz} \\ \tau_{xy} & \sigma_{yy} & \tau_{yz} \\ \tau_{xz} & \tau_{yz} & \sigma_{zz} \end{bmatrix}_{(\underline{x},\underline{y},\underline{z})} \qquad [1.30]$$

The very notion itself of the stress matrix is important, as it assumes that at one point, and in the small volume (REV) surrounding it, the state of the internal forces of matter is entirely represented by this matrix. This is in contrast to the stress vector, which only gives the internal force for one single facet.

EXAMPLE: DRAWING THE STRESSES IN 2D.–

A small square is subjected to a 2D-stress tensor:

$$\underline{\underline{\sigma}}(M) = \begin{bmatrix} \sigma_{xx} & \tau_{xy} \\ \tau_{xy} & \sigma_{yy} \end{bmatrix}_{(\underline{x},\underline{y})} \qquad [1.31]$$

These faces are therefore subjected to the following forces:

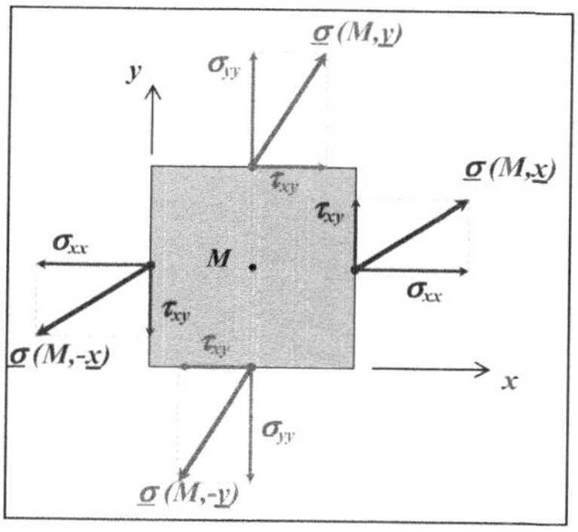

Figure 1.12. *Stress vectors on the faces of a square. For a color version of this figure, see www.iste.co.uk/bouvet/aeronautical.zip*

This drawing may be simple, but it is paramount for interpreting the stress tensor.

You obviously would have noticed that the projection of $\underline{\sigma}(M, \underline{x})$ on $\underline{y}$ is equal to that of $\underline{\sigma}(M, \underline{y})$ on $\underline{x}$, which is evidently due to the symmetry of the stress tensor.

I will leave it up to you to do the same drawing in 3D.

1.3.2. *Invariants of the stress tensor*

The stress tensor possesses three elementary invariants. We classically use:

– The hydrostatic pressure:

$$p = \frac{1}{3} trace\left(\underline{\underline{\sigma}} \right) = \frac{\sigma_x + \sigma_y + \sigma_z}{3} \quad [1.32]$$

It is named thus because when we apply a uniform pressure to a cube in all directions, we obtain:

$$\underline{\underline{\sigma}} = \begin{bmatrix} -p & 0 & 0 \\ 0 & -p & 0 \\ 0 & 0 & -p \end{bmatrix} \quad [1.33]$$

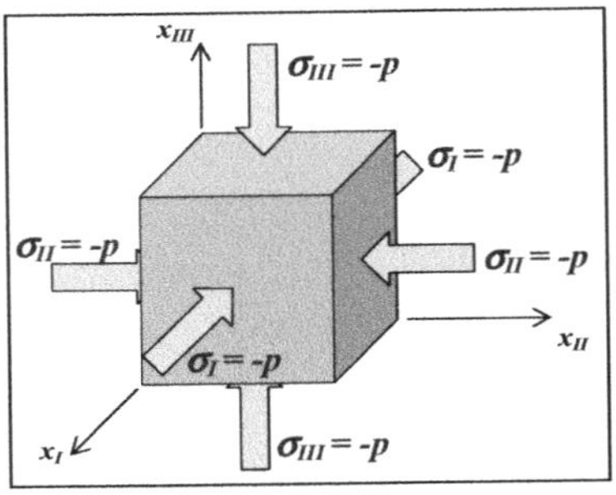

Figure 1.13. *Hydrostatic pressure. For a color version of this figure, see www.iste.co.uk/bouvet/aeronautical.zip*

In this case, the hydrostatic pressure is evidently equal to the pressure p, in its absolute value.

– The Von Mises equivalent stress:

$$\overline{\sigma} = \sqrt{\frac{3}{2}.\underline{dev}\left(\underline{\underline{\sigma}}\right):\underline{dev}\left(\underline{\underline{\sigma}}\right)} = \sqrt{\frac{1}{2}.\left((\sigma_I - \sigma_{II})^2 + (\sigma_{II} - \sigma_{III})^2 + (\sigma_{III} - \sigma_I)^2\right)} \quad [1.34]$$

With the stress deviator, which is written as:

$$\underline{dev}\left(\underline{\underline{\sigma}}\right) = \underline{\underline{\sigma}} - \frac{1}{3}.trace\left(\underline{\underline{\sigma}}\right).\underline{\underline{I}} \quad [1.35]$$

This stress is very important for estimating the beginning of a ductile material's plasticity (see Chapters 3 and 4).

– The determinant:

$$I_3 = \det\left(\underline{\underline{\sigma}}\right) \quad [1.36]$$

These invariants are very important for writing the fracture, yield or damage criteria. Given that a criterion of this type is indeed representative of the state of matter, it must not depend on the coordinate at which we write the stress matrix, and therefore it can be written based on these invariants.

1.3.3. *Relation between the stress matrix and the stress vector*

By constructing the stress matrix, we then have:

$$\begin{cases} \underline{\sigma}(M,\underline{x}) = \underline{\underline{\sigma}}(M).\underline{x} \\ \underline{\sigma}(M,\underline{y}) = \underline{\underline{\sigma}}(M).\underline{y} \\ \underline{\sigma}(M,\underline{z}) = \underline{\underline{\sigma}}(M).\underline{z} \end{cases} \quad [1.37]$$

And so, irrespective of $\underline{n}$:

$$\underline{\sigma}(M,\underline{n}) = \underline{\underline{\sigma}}(M).\underline{n} \quad [1.38]$$

This relation is very important as it enables the stress matrix to be linked to the stress vector. It will therefore be used very often throughout the rest of this book.

EXAMPLE: TRACTION.–

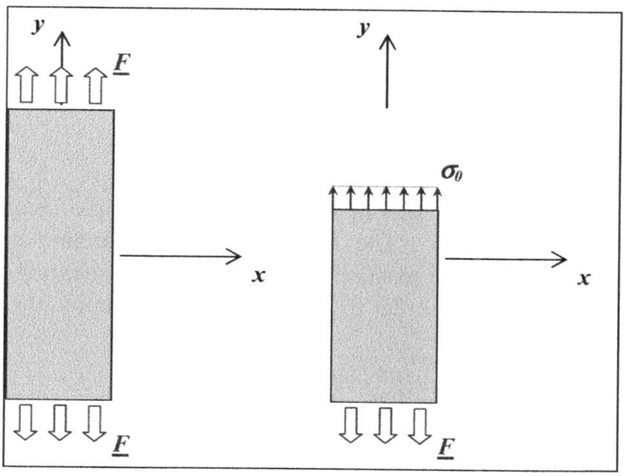

Figure 1.14. *Traction test. For a color version of this figure, see www.iste.co.uk/bouvet/aeronautical.zip*

Let us search for the stress tensor of a traction test. The boundary conditions on the six external faces give:

$$\begin{cases} \underline{\sigma}(M,\underline{x}) = \underline{\underline{\sigma}}(M).\underline{x} = \underline{0} \\ \underline{\sigma}(M,-\underline{x}) = \underline{\underline{\sigma}}(M).(-\underline{x}) = \underline{0} \\ \iint_{S_y} \underline{\sigma}(M,\underline{y}).dS = \iint_{S_y} \underline{\underline{\sigma}}(M).\underline{y}.dS = \underline{F} = F.\underline{y} \\ \iint_{S_y} \underline{\sigma}(M,-\underline{y}).dS = \iint_{S_y} \underline{\underline{\sigma}}(M).(-\underline{y}).dS = -\underline{F} \\ \underline{\sigma}(M,\underline{z}) = \underline{\underline{\sigma}}(M).\underline{z} = \underline{0} \\ \underline{\sigma}(M,-\underline{z}) = \underline{\underline{\sigma}}(M).(-\underline{z}) = \underline{0} \end{cases} \quad [1.39]$$

Moreover, if we assume the stresses to be homogeneous:

$$\underline{\sigma}(M,\underline{y}) = \underline{\underline{\sigma}}(M).\underline{y} = \frac{\underline{F}}{S} = \sigma_0.\underline{y} \quad [1.40]$$

Hence, the classic result:

$$\underline{\underline{\sigma}}(M) = \begin{bmatrix} 0 & 0 & 0 \\ 0 & \sigma_0 & 0 \\ 0 & 0 & 0 \end{bmatrix}_{(\underline{x},\underline{y},\underline{z})} \quad [1.41]$$

Experimentally, during a traction test on a metal sample, we note that the fractography is oriented at 45°. The fracture obviously occurs after significant plasticity. Yet we know that the plasticity is sensitive to the shearing, rather than to the normal stress. Therefore, in order to explain this fractography, we can show that the shearing is maximum at 45°.

This exercise can be done in 2D, or even in 3D, as the demonstration will be similar:

$$\underline{\underline{\sigma}}(M) = \begin{bmatrix} 0 & 0 \\ 0 & \sigma_0 \end{bmatrix}_{(\underline{x},\underline{y})} \quad [1.42]$$

And for a facet with normal $\underline{n}$, the normal and shear stresses are:

$$\underline{\sigma}(M,\underline{n}) = \underline{\underline{\sigma}}(M).\underline{n} = \sigma_0 . \sin(\theta).\underline{y} = \sigma_{nn}.\underline{n} + \tau_{nt}.\underline{t} \quad [1.43]$$

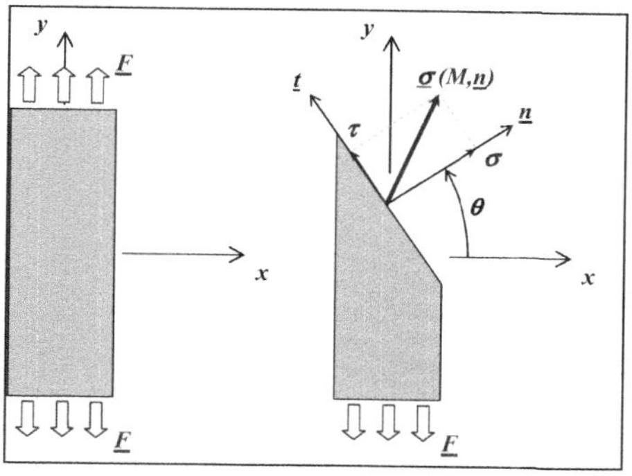

Figure 1.15. *Stress vectors in traction. For a color version of this figure, see www.iste.co.uk/bouvet/aeronautical.zip*

where:

$$\begin{cases} \sigma_{nn} = \sigma_0 \, .\sin^2(\theta) \\ \tau_{nt} = \sigma_0 \, .\sin(\theta).\cos(\theta) \end{cases} \quad [1.44]$$

So, τ is maximum at 45°.

Calculation using the equilibrium of a triangle of matter:

Knowing that the stress matrix is built on the basis of the equilibrium of a REV, the previous result can be found again by balancing a triangle of matter.

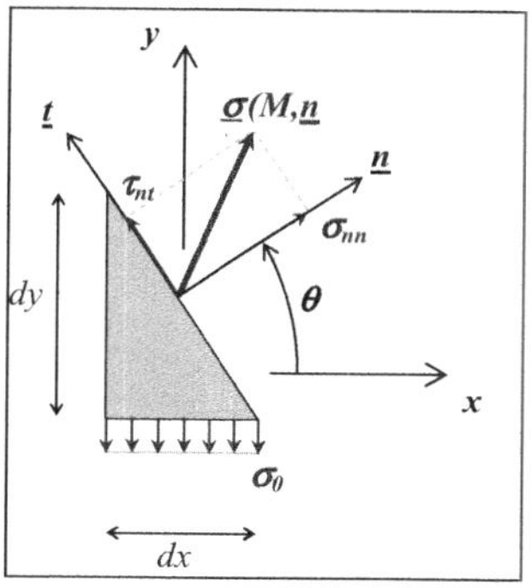

Figure 1.16. *Equilibrium of a triangle of matter. For a color version of this figure, see www.iste.co.uk/bouvet/aeronautical.zip*

Equilibrium in accordance with $\underline{x}$:

$$\sigma_{nn} . \frac{dx}{\sin(\theta)} . \cos(\theta) - \tau_{nt} . \frac{dx}{\sin(\theta)} . \sin(\theta) = 0 \qquad [1.45]$$

Equilibrium according to $\underline{y}$:

$$-\sigma_0 . dx + \sigma_{nn} . \frac{dx}{\sin(\theta)} . \sin(\theta) + \tau_{nt} . \frac{dx}{\sin(\theta)} . \cos(\theta) = 0 \qquad [1.46]$$

where:

$$\begin{cases} \sigma_{nn} = \sigma_0 . \sin^2(\theta) \\ \tau_{nt} = \sigma_0 . \sin(\theta) . \cos(\theta) \end{cases} \qquad [1.47]$$

We then evidently find the previous result again.

1.3.4. *Principal stresses and principal directions*

As the 3D stress matrix is:

$$\underline{\underline{\sigma}}(M) = \begin{bmatrix} \sigma_{xx} & \tau_{xy} & \tau_{xz} \\ \tau_{xy} & \sigma_{yy} & \tau_{yz} \\ \tau_{xz} & \tau_{yz} & \sigma_{zz} \end{bmatrix}_{(\underline{x},\underline{y},\underline{z})} \qquad [1.48]$$

Evidently, this stress matrix can be determined in another coordinate system $(\underline{x}_1, \underline{y}_1, \underline{z}_1)$ by:

$$\underline{\underline{\sigma}}(M)_{B_1} = \underline{\underline{P}}^t . \underline{\underline{\sigma}}(M)_B . \underline{\underline{P}} \qquad [1.49]$$

with $\underline{\underline{P}}$, the rotation matrix from the basis B to the basis B_1, representing the vector coordinates of B_1 expressed in the basis B.

EXAMPLE: TRACTION.–

Thanks to this relation, we can evidently find the normal and shear stresses of a traction tensor:

$$\underline{\underline{\sigma}}(M)=\begin{bmatrix}0&0&0\\0&\sigma_0&0\\0&0&0\end{bmatrix}_{(\underline{x},\underline{y},\underline{z})}=\begin{bmatrix}\sin^2(\theta).\sigma_0&\cos(\theta).\sin(\theta).\sigma_0&0\\\cos(\theta).\sin(\theta).\sigma_0&\cos^2(\theta).\sigma_0&0\\0&0&0\end{bmatrix}_{(\underline{n},\underline{t},\underline{z})} \quad [1.50]$$

With:

$$\underline{\underline{P}}=\begin{bmatrix}\cos(\theta)&-\sin(\theta)&0\\\sin(\theta)&\cos(\theta)&0\\0&0&1\end{bmatrix}_{(\underline{x},\underline{y},\underline{z})} \quad [1.51]$$

And in the stress matrix expressed in B_1, we will recognize the expressions of σ_{nn} and σ_{nt}.

THEOREM.– *There is a direct orthonormal coordinate system ($\underline{x}_I$, $\underline{x}_{II}$, $\underline{x}_{III}$) in which the stress matrix is diagonal*:

$$\underline{\underline{\sigma}}(M)=\begin{bmatrix}\sigma_I&0&0\\0&\sigma_{II}&0\\0&0&\sigma_{III}\end{bmatrix}_{(\underline{x}_I,\underline{x}_{II},\underline{x}_{III})} \quad [1.52]$$

σ_I, σ_{II} and σ_{III} are called principal stresses and $\underline{x}_I$, $\underline{x}_{II}$ and $\underline{x}_{III}$ are principal directions (or eigenvectors for mathematicians) associated with σ_I, σ_{II} and σ_{III} respectively.

We then clearly have:

$$\begin{cases}\underline{\sigma}(M,\underline{x}_I)=\underline{\underline{\sigma}}(M).\underline{x}_I=\sigma_I.\underline{x}_I\\\underline{\sigma}(M,\underline{x}_{II})=\underline{\underline{\sigma}}(M).\underline{x}_{II}=\sigma_{II}.\underline{x}_{II}\\\underline{\sigma}(M,\underline{x}_{III})=\underline{\underline{\sigma}}(M).\underline{x}_{III}=\sigma_{III}.\underline{x}_{III}\end{cases} \quad [1.53]$$

In practice, to determine the principal stresses, it is sufficient to write the below:

$$\det\left(\underline{\underline{\sigma}}(M)-\sigma_i.\underline{\underline{I}}\right)=0 \qquad [1.54]$$

which gives three solutions (or two in 2D). Then, to determine the three principal directions, it is sufficient to write the three previous relations. In practice, the three principal directions are orthogonal, so when two of them have been found, the third one can be deduced from the other two.

This can be translated by the following drawing:

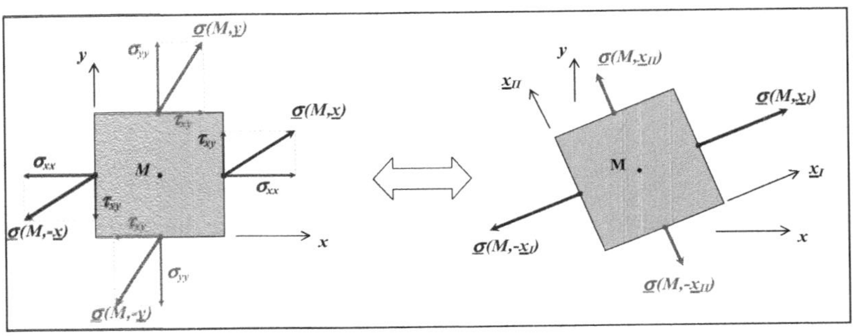

Figure 1.17. *Principal stresses and associated stress vectors. For a color version of this figure, see www.iste.co.uk/bouvet/aeronautical.zip*

The state of the stress as seen by the matter on these two diagrams is the same.

EXAMPLE: SHEARING.–

As the stress tensor is:

$$\underline{\underline{\sigma}}(M)=\begin{bmatrix}0 & \tau\\ \tau & 0\end{bmatrix}_{(\underline{x},\underline{y})} \qquad [1.55]$$

So, we will be able to demonstrate that the principal stresses are $+\tau$ and $-\tau$, and the principal directions are oriented at +45° and –45°:

On a physical level, this result is easy to understand as we can easily use our hands to feel that the applied force on the left diagram pulls at +45° and compresses at –45°.

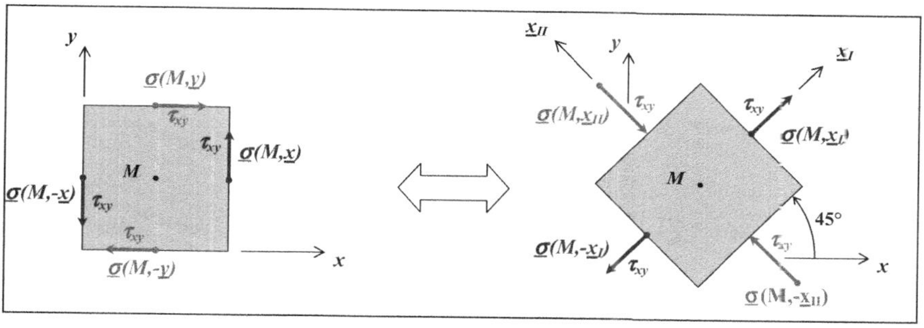

Figure 1.18. *Stress vectors for pure shearing. For a color version of this figure, see www.iste.co.uk/bouvet/aeronautical.zip*

1.4. Equilibrium equation

This exercise is to be done in 2D:

$$\underline{\underline{\sigma}}(M)=\begin{bmatrix}\sigma_{xx} & \tau_{xy}\\ \tau_{xy} & \sigma_{yy}\end{bmatrix}_{(\underline{x},\underline{y})} \qquad [1.56]$$

And now, we shall study stress vectors on a small square:

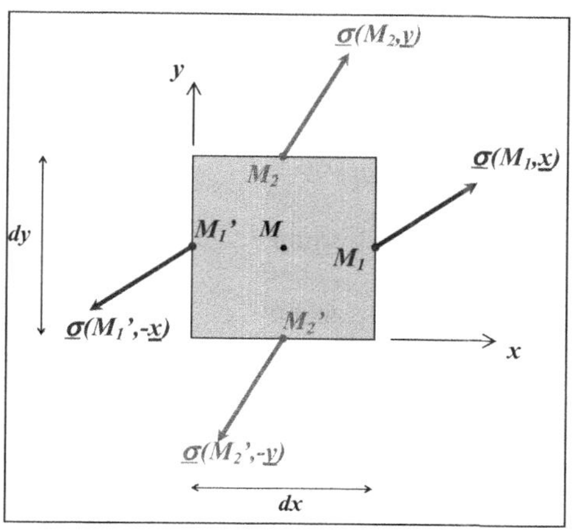

Figure 1.19. *Stress vectors on the faces of a square. For a color version of this figure, see www.iste.co.uk/bouvet/aeronautical.zip*

This square is subjected to the following stress vectors:

$$\begin{cases} \underline{\sigma}(M_1,\underline{x}) = \underline{\underline{\sigma}}(M_1).\underline{x} = \left(\underline{\underline{\sigma}}(M) + \dfrac{\partial \underline{\underline{\sigma}}(M)}{\partial x}.\dfrac{dx}{2}\right).\underline{x} \\ \underline{\sigma}(M_1',-\underline{x}) = \underline{\underline{\sigma}}(M_1').(-\underline{x}) = \left(\underline{\underline{\sigma}}(M) - \dfrac{\partial \underline{\underline{\sigma}}(M)}{\partial x}.\dfrac{dx}{2}\right).(-\underline{x}) \\ \underline{\sigma}(M_2,\underline{y}) = \underline{\underline{\sigma}}(M_2).\underline{y} = \left(\underline{\underline{\sigma}}(M) + \dfrac{\partial \underline{\underline{\sigma}}(M)}{\partial y}.\dfrac{dy}{2}\right).\underline{y} \\ \underline{\sigma}(M_2',-\underline{y}) = \underline{\underline{\sigma}}(M_2').(-\underline{y}) = \left(\underline{\underline{\sigma}}(M) - \dfrac{\partial \underline{\underline{\sigma}}(M)}{\partial y}.\dfrac{dy}{2}\right).(-\underline{y}) \end{cases} \quad [1.57]$$

If we assume that the cube is subjected to a volume force $\underline{f}_v$, then the equilibrium equation in force can be written as:

$$\underline{\sigma}(M_1,\underline{x}).dy.dz + \underline{\sigma}(M_1',-\underline{x}).dy.dz + \underline{\sigma}(M_2,\underline{y}).dx.dz \\ + \underline{\sigma}(M_2',-\underline{y}).dx.dz + \underline{f}_v.dx.dy.dz = \underline{0} \quad [1.58]$$

Hence:

$$\begin{cases} \dfrac{\partial \sigma_{xx}}{\partial x} + \dfrac{\partial \tau_{xy}}{\partial y} + f_{vx} = 0 \\ \dfrac{\partial \tau_{xy}}{\partial x} + \dfrac{\partial \sigma_{yy}}{\partial y} + f_{vy} = 0 \end{cases} \quad [1.59]$$

This equation represents the equilibrium equation of the cube.

Using the same approach in 3D leads to:

$$\begin{cases} \dfrac{\partial \sigma_{xx}}{\partial x} + \dfrac{\partial \tau_{xy}}{\partial y} + \dfrac{\partial \tau_{xz}}{\partial z} + f_{vx} = 0 \\ \dfrac{\partial \tau_{xy}}{\partial x} + \dfrac{\partial \sigma_{yy}}{\partial y} + \dfrac{\partial \tau_{yz}}{\partial z} + f_{vy} = 0 \\ \dfrac{\partial \tau_{xz}}{\partial x} + \dfrac{\partial \tau_{yz}}{\partial y} + \dfrac{\partial \sigma_{zz}}{\partial z} + f_{vz} = 0 \end{cases} \quad [1.60]$$

This equation can also be written as:

$$\underline{div}\left(\underline{\underline{\sigma}}\right)+\underline{f}_v=\underline{0} \qquad [1.61]$$

This is a fundamental equation as it alone translates the equilibrium of a material point, and therefore also that of the small volume surrounding it. It must therefore be verified at each point of the solid. This is one of the fundamental equations which are needed to verify a stress field (see Chapter 3) in order to be the solution to the problem.

1.5. Mohr's circle

Mohr's circle is a graphic method used to rotate the matrix in 2D. It enables us to simply determine the maximum shear stresses, the normal stresses, the principal stresses, the principal directions, etc.

As the 2D stress vector is:

$$\underline{\underline{\sigma}}(M)=\begin{bmatrix}\sigma_{xx} & \tau_{xy}\\ \tau_{xy} & \sigma_{yy}\end{bmatrix}_{(\underline{x},\underline{y})} \qquad [1.62]$$

Mohr's circle is a group of points whose abscissa is a normal stress σ and whose ordinate is the shear stress τ for all the possible facets:

As $\underline{n}$ is a vector of the plane, then:

$$P\begin{bmatrix}\sigma\\ \tau\end{bmatrix} \qquad [1.63]$$

with:

$$\begin{cases}\sigma=\underline{\sigma}(P,\underline{n}).\underline{n}\\ \tau=\underline{\sigma}(P,\underline{n}).\underline{t}\end{cases} \qquad [1.64]$$

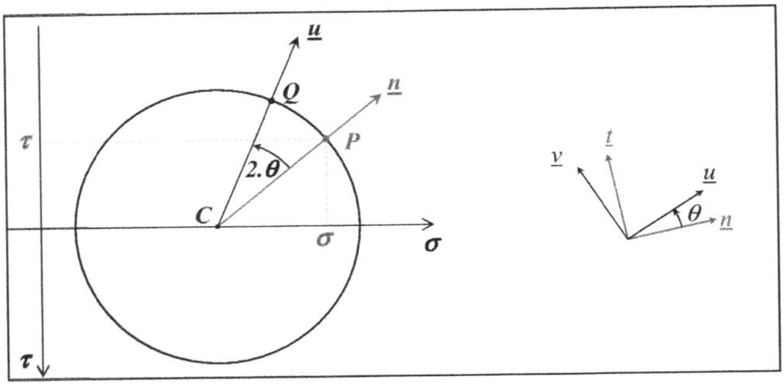

Figure 1.20. *Mohr's circle. For a color version of this figure, see www.iste.co.uk/bouvet/aeronautical.zip*

The $\underline{CP}$ vector is thus representative of the facet with normal $\underline{n}$, and when $\underline{n}$ varies, the point P describes a circle with a center C located on the abscissa axis.

It can also be shown that while passing from the vector $\underline{n}$ to vector $\underline{u}$, making an angle θ with $\underline{n}$, then an angle $2.\theta$ is made within Mohr's circle (this incidentally ensures that this angle remains equal to $+2.\theta$, and not to $-2.\theta$, when we take the downwards-facing shearing axis).

Knowing the stress tensor is:

$$\underline{\underline{\sigma}}(M) = \begin{bmatrix} \sigma_{xx} & \tau_{xy} \\ \tau_{xy} & \sigma_{yy} \end{bmatrix}_{(\underline{x},\underline{y})} \quad [1.65]$$

We can then trace the points X (σ_{xx}, τ_{xy}) and Y (σ_{yy}, $-\tau_{xy}$), and trace the center C and Mohr's circle. From this, we can therefore deduce the main stresses σ_I and σ_{II} and the principal directions.

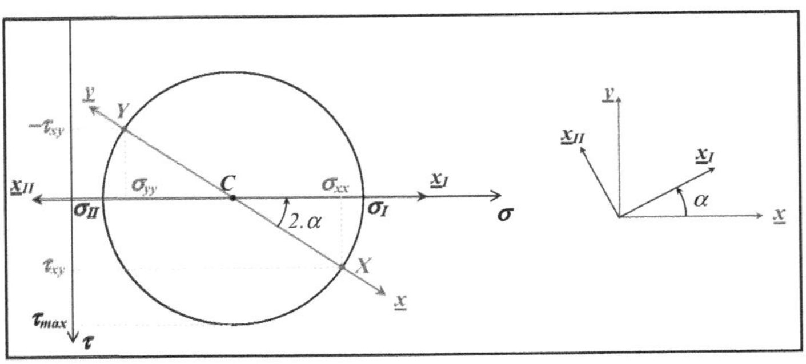

Figure 1.21. *Mohr's circle. For a color version of this figure, see www.iste.co.uk/bouvet/aeronautical.zip*

From Mohr's circle, we can also deduce that the maximum normal stress is either σ_I or σ_{II} and that the maximum shearing $\tau_{\max}$ is equal to:

$$\tau_{\max} = \frac{|\sigma_I - \sigma_{II}|}{2} \quad [1.66]$$

This relation can moreover be generalized in 3D:

$$\tau_{\max} = Max\left(\frac{|\sigma_I - \sigma_{II}|}{2};\frac{|\sigma_{II} - \sigma_{III}|}{2};\frac{|\sigma_{III} - \sigma_I|}{2}\right) \quad [1.67]$$

We can also show that in 2D:

$$\tau_{\max} = \sqrt{\left(\frac{\sigma_{xx} - \sigma_{yy}}{2}\right)^2 + \tau_{xy}{}^2} \quad [1.68]$$

$$\sigma_{\max} = Max\left(\frac{\sigma_{xx} + \sigma_{yy}}{2} + \sqrt{\left(\frac{\sigma_{xx} - \sigma_{yy}}{2}\right)^2 + \tau_{xy}{}^2} ; \frac{\sigma_{xx} + \sigma_{yy}}{2} - \sqrt{\left(\frac{\sigma_{xx} - \sigma_{yy}}{2}\right)^2 + \tau_{xy}{}^2}\right) \quad [1.69]$$

EXAMPLE: TRACTION.–

The stress tensor in traction is:

$$\underline{\underline{\sigma}}(M) = \begin{bmatrix} 0 & 0 \\ 0 & \sigma_0 \end{bmatrix}_{(\underline{x},\underline{y})} \quad [1.70]$$

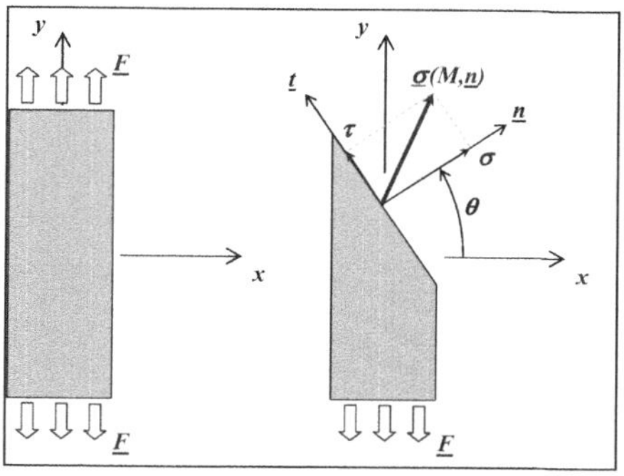

Figure 1.22. *Stress vectors on a facet with a normal vector n in traction. For a color version of this figure, see www.iste.co.uk/bouvet/aeronautical.zip*

Mohr's circle can then be traced:

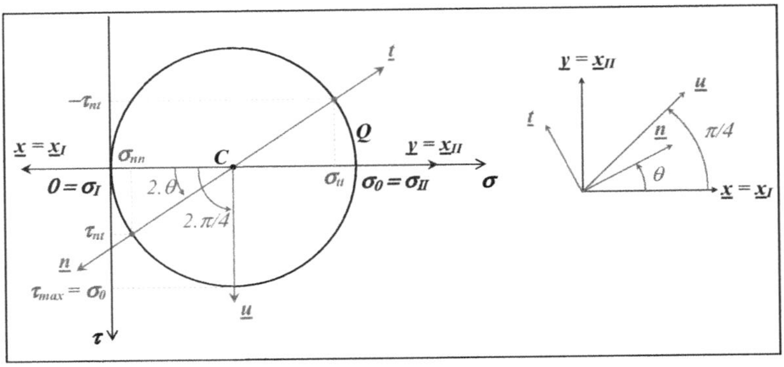

Figure 1.23. *Mohr's circle in traction. For a color version of this figure, see www.iste.co.uk/bouvet/aeronautical.zip*

We then find:

$$\begin{cases} \sigma = \sigma_{nn} = \dfrac{\sigma_0}{2} - \dfrac{\sigma_0}{2}.\cos(2.\theta) = \sigma_0.\sin^2(\theta) \\ \tau = \tau_{nt} = \dfrac{\sigma_0}{2}.\sin(2.\theta) = \sigma_0.\sin(\theta).\cos(\theta) \end{cases} \quad [1.71]$$

The maximum shearing at 45° is equal to σ_0.

2

Strain

2.1. Notion of strain

2.1.1. *Displacement vector*

Let us say that S is a solid and $M(x, y, z)$ is a point of S. Under the action of external forces, S becomes S' and M becomes M'. This is called the displacement vector of the point M, and we note $\underline{u}(M)$, the vector $\underline{MM'}$.

$$\underline{u}(M) = \underline{MM}' = \begin{vmatrix} u(x, y, z) \\ v(x, y, z) \\ w(x, y, z) \end{vmatrix} \qquad [2.1]$$

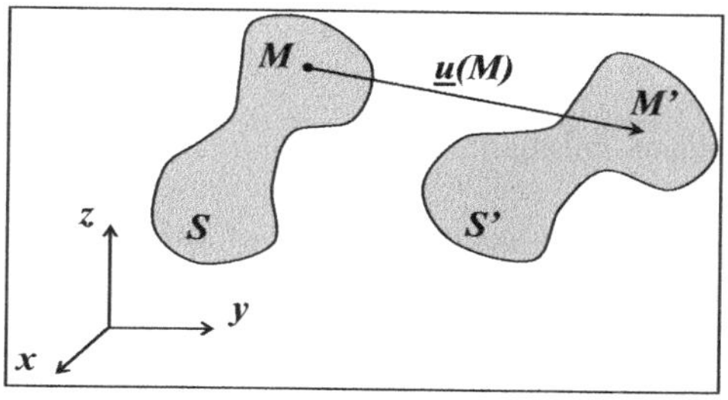

Figure 2.1. *Displacement field of a solid*

2.1.2. *Unit strain*

Thus, there are two points of *S*: *M* and *N*, which are displaced to *M'* and *N'* after stress and $\underline{n}$ the unit vector:

$$\underline{n} = \frac{\underline{\underline{MN}}}{MN} \qquad [2.2]$$

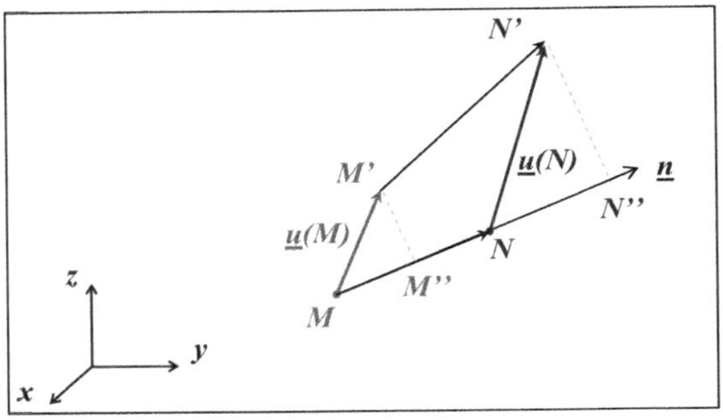

Figure 2.2. *Unit strain. For a color version of this figure, see www.iste.co.uk/bouvet/aeronautical.zip*

This is called unit strain in *M* according to $\underline{n}$:

$$\varepsilon\left(M,\underline{n}\right) = \varepsilon_n(M) = \lim_{N \to M}\left(\frac{\underline{\underline{M'N'}}.\underline{n} - MN}{MN}\right) \qquad [2.3]$$

Evidently, $\varepsilon(M,\underline{n})$ has no unit.

This strain $\varepsilon(M,\underline{n})$ must remain small in front of one in order to belong within the Small Perturbation Hypothesis (SPH) .

As indicated by its name, this hypothesis translates the length strain of the matter in the *n*-direction.

In particular, if $\underline{n}$ is equal to $\underline{x}$, and the relation is put in 2D in order to facilitate representations:

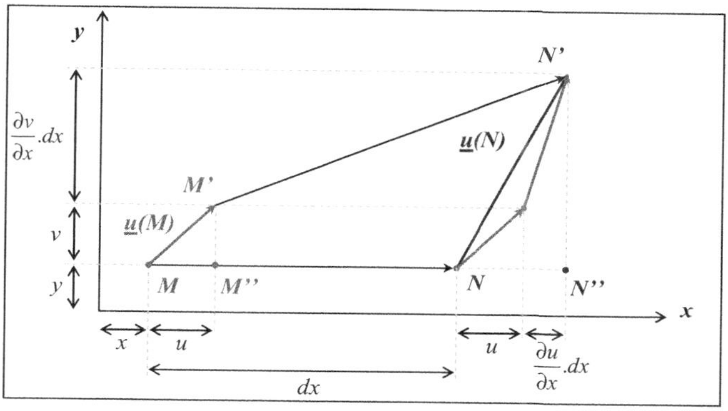

Figure 2.3. *Unit strain formula. For a color version of this figure, see www.iste.co.uk/bouvet/aeronautical.zip*

As $N(x + dx, y)$ is close to $M(x, y)$, we have:

$$\underline{u}(M) = \underline{MM'} = \begin{vmatrix} u(x, y) \\ v(x, y) \end{vmatrix} \quad [2.4]$$

And

$$\underline{u}(N) = \underline{NN'} = \begin{vmatrix} u(x+dx, y) \\ v(x+dx, y) \end{vmatrix} = \begin{vmatrix} u(x, y) + \dfrac{\partial u}{\partial x}(x, y).dx \\ v(x, y) + \dfrac{\partial v}{\partial x}(x, y).dx \end{vmatrix} \quad [2.5]$$

Hence:

$$\varepsilon\left(M, \underline{x}\right) = \varepsilon_x(M) = \lim_{dx \to 0}\left(\frac{M''N'' - MN}{MN}\right) = \frac{\partial u}{\partial x}(x, y) \quad [2.6]$$

Evidently in 3D, we can make the following displacement generalization for a point $M(x, y, z)$:

$$\underline{u}(M) = \begin{vmatrix} u(x, y, z) \\ v(x, y, z) \\ w(x, y, z) \end{vmatrix} \quad [2.7]$$

These strains are defined according to the x-, y- and z- directions by:

$$\begin{cases} \varepsilon\left(M,\underline{x}\right)=\varepsilon_x(M)=\dfrac{\partial u}{\partial x}(x,y,z) \\ \varepsilon\left(M,\underline{y}\right)=\varepsilon_y(M)=\dfrac{\partial v}{\partial y}(x,y,z) \\ \varepsilon\left(M,\underline{z}\right)=\varepsilon_z(M)=\dfrac{\partial w}{\partial z}(x,y,z) \end{cases} \qquad [2.8]$$

And ε_x, ε_y and ε_z therefore translate the length strain of the matter in the x-, y-, and z-directions, respectively.

2.1.3. *Angular distortion*

As defined by the figure above, M is one point and M_1 and M_2 are two points that are close to M, which are displaced to M', M_1' and M_2' respectively.

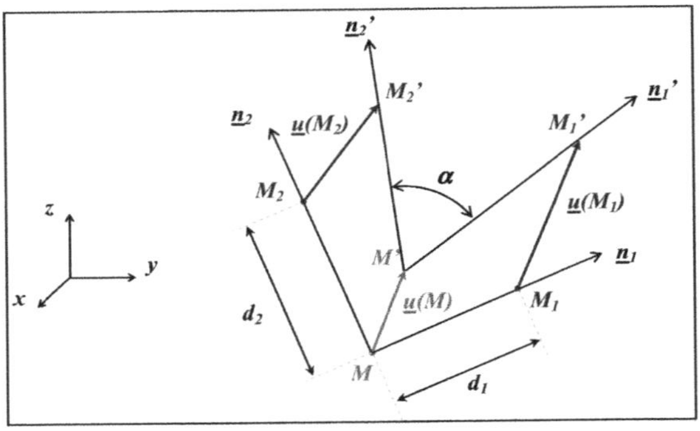

Figure 2.4. *Angular distortion. For a color version of this figure, see www.iste.co.uk/bouvet/aeronautical.zip*

The two unit vectors are defined as:

$$\begin{cases} \underline{n}_1=\dfrac{\underline{MM_1}}{MM_1} \\ \underline{n}_2=\dfrac{\underline{MM_2}}{MM_2} \end{cases} \qquad [2.9]$$

According to $\underline{n}_1$ and $\underline{n}_2$, the angular distortion is at M:

$$\gamma(M,\underline{n}_1,\underline{n}_2)=\gamma_{n_1n_2}(M)=\lim_{\substack{d_1\to0\\d_2\to0}}\left(\frac{\pi}{2}-\alpha\right) \qquad [2.10]$$

Evidently, $\gamma(M, \underline{n}_1, \underline{n}_2)$ has no unit.

This strain $\gamma(M, \underline{n}_1, \underline{n}_2)$ must remain small in front of one in order to belong within the SPH.

It translates the angle variation of the matter, called shear, of the angle that was initially a right-angled corner $(\underline{n}_1, \underline{n}_2)$. Altogether, the larger the distortion, less right-angled the corner will be.

In particular, if $\underline{n}_1$ is equal to $\underline{x}$, if $\underline{n}_2$ is equal to $\underline{y}$, and placed in 2D in order to facilitate the representations:

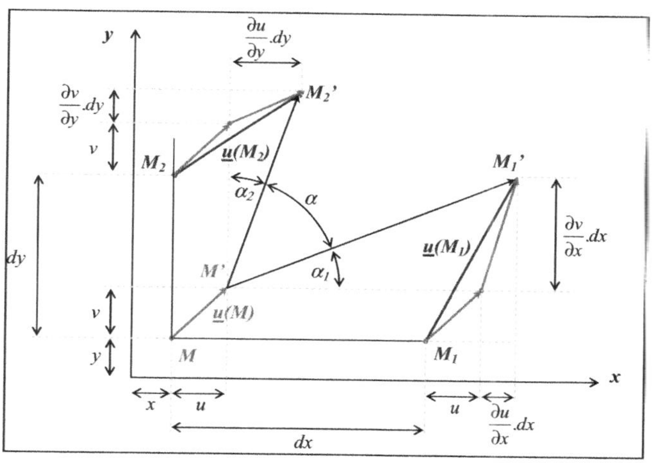

Figure 2.5. *Angular distortion formula. For a color version of this figure, see www.iste.co.uk/bouvet/aeronautical.zip*

As $M_1(x + dx, y)$ and $M_2(x, y + dy)$ are close to $M(x, y)$, we have:

$$\underline{u}(M)=\underline{MM}'=\begin{vmatrix}u(x,y)\\v(x,y)\end{vmatrix} \qquad [2.11]$$

And:

$$\begin{cases} \underline{u}(M_1) = \underline{M_1M_1'} = \begin{vmatrix} u(x+dx,y) \\ v(x+dx,y) \end{vmatrix} = \begin{vmatrix} u(x,y)+\dfrac{\partial u}{\partial x}(x,y).dx \\ v(x,y)+\dfrac{\partial v}{\partial x}(x,y).dx \end{vmatrix} \\ \underline{u}(M_2) = \underline{M_2M_2'} = \begin{vmatrix} u(x,y+dy) \\ v(x,y+dy) \end{vmatrix} = \begin{vmatrix} u(x,y)+\dfrac{\partial u}{\partial y}(x,y).dy \\ v(x,y)+\dfrac{\partial v}{\partial y}(x,y).dy \end{vmatrix} \end{cases} \quad [2.12]$$

Yet:

$$\alpha = \left(\underline{M'M_1'};\underline{M'M_2'}\right) = \frac{\pi}{2} - \alpha_1 - \alpha_2 \quad [2.13]$$

Hence:

$$\begin{cases} \alpha_1 = \dfrac{\dfrac{\partial v}{\partial x}.dx}{dx+\dfrac{\partial u}{\partial x}.dx} \approx \dfrac{\partial v}{\partial x} \\ \alpha_2 = \dfrac{\dfrac{\partial u}{\partial y}.dy}{dy+\dfrac{\partial u}{\partial y}.dy} \approx \dfrac{\partial u}{\partial y} \end{cases} \quad [2.14]$$

And:

$$\gamma\left(M,\underline{x},\underline{y}\right) = \gamma_{xy}(M) = \frac{\partial u}{\partial y} + \frac{\partial v}{\partial x} \quad [2.15]$$

Evidently in 3D, we can make a generalization for point $M(x, y, z)$, for which the displacement is worth:

$$\underline{u}(M) = \begin{vmatrix} u(x,y,z) \\ v(x,y,z) \\ w(x,y,z) \end{vmatrix} \quad [2.16]$$

The angular distortions are therefore defined by:

$$\begin{cases} \gamma_{xy}(M) = \gamma_{yx}(M) = \dfrac{\partial u}{\partial y} + \dfrac{\partial v}{\partial x} \\ \gamma_{yz}(M) = \gamma_{zy}(M) = \dfrac{\partial v}{\partial z} + \dfrac{\partial w}{\partial y} \\ \gamma_{xz}(M) = \gamma_{zx}(M) = \dfrac{\partial w}{\partial x} + \dfrac{\partial u}{\partial z} \end{cases} \qquad [2.17]$$

Therefore, (with i and j as variants of 1–3, and i being different to j) we evidently have:

$$\gamma_{ij} = \gamma_{ji} \qquad [2.18]$$

γ_{xy}, γ_{yz} and γ_{xz} therefore translate the angle variation of the matter for the angles which were initially right-angled corners: $(\underline{x}, \underline{y})$, $(\underline{y}, \underline{z})$ and $(\underline{z}, \underline{x})$, respectively.

2.2. Strain matrix

2.2.1. *Definition of the strain matrix*

The following matrix is called the strain matrix $\underline{\underline{\varepsilon}}(M)$:

$$\underline{\underline{\varepsilon}}(M) = \begin{bmatrix} \varepsilon_x & \varepsilon_{xy} & \varepsilon_{xz} \\ \varepsilon_{xy} & \varepsilon_y & \varepsilon_{yz} \\ \varepsilon_{xz} & \varepsilon_{yz} & \varepsilon_z \end{bmatrix}_{(\underline{x},\underline{y},\underline{z})} \qquad [2.19]$$

With (be careful here):

$$\begin{cases} \varepsilon_{xy}(M) = \dfrac{\gamma_{xy}(M)}{2} \\ \varepsilon_{yz}(M) = \dfrac{\gamma_{yz}(M)}{2} \\ \varepsilon_{xz}(M) = \dfrac{\gamma_{xz}(M)}{2} \end{cases} \qquad [2.20]$$

This matrix is therefore obviously symmetrical.

Based on the displacement field, it is defined by the relations seen previously:

$$\begin{cases} \varepsilon_x = \dfrac{\partial u}{\partial x} \\ \varepsilon_y = \dfrac{\partial v}{\partial y} \\ \varepsilon_z = \dfrac{\partial w}{\partial z} \\ 2.\varepsilon_{xy} = \gamma_{xy} = \dfrac{\partial u}{\partial y} + \dfrac{\partial v}{\partial x} \\ 2.\varepsilon_{yz} = \gamma_{yz} = \dfrac{\partial v}{\partial z} + \dfrac{\partial w}{\partial y} \\ 2.\varepsilon_{xz} = \gamma_{xz} = \dfrac{\partial w}{\partial x} + \dfrac{\partial u}{\partial z} \end{cases} \quad [2.21]$$

Or, in the tensorial form:

$$\underline{\underline{\varepsilon}} = \frac{1}{2}.\left(\underline{\underline{grad}}(\underline{u}) + \underline{\underline{grad}}^t(\underline{u}) \right) \quad [2.22]$$

The benefit of using the tensorial form is of course that it remains true for every coordinate system (Cartesian, cylindrical or spherical), in contrast to the previous form, which is only true in Cartesian coordinates. All the same, the displacement gradient of the coordinate system in question remains to be determined.

EXAMPLES: DRAWING THE 2D STRAINS.–

A small square is subjected to the strain tensor:

$$\underline{\underline{\varepsilon}}(M) = \begin{bmatrix} \varepsilon_x & \varepsilon_{xy} \\ \varepsilon_{xy} & \varepsilon_y \end{bmatrix}_{(\underline{x},\underline{y})} \quad [2.23]$$

It is strained in the following way:

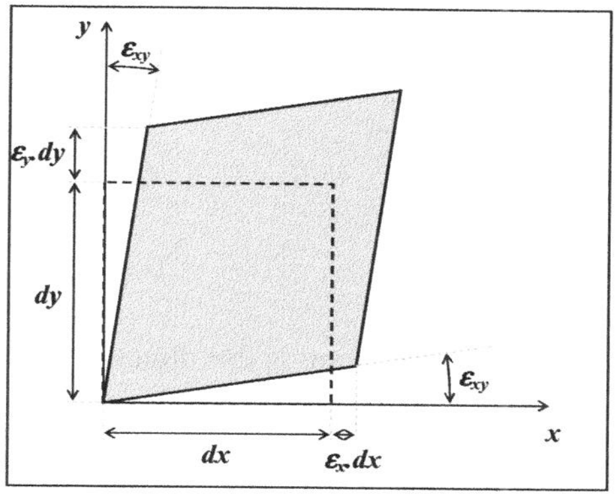

Figure 2.6. *Strain of a square*

This drawing may be simple and basic but it is essential to interpret the strain tensor. Moreover, it is only valid for SPH.

Finally, it is defined at a close translation and rotation. Indeed, two displacement fields, which differ from a rigid body displacement field, produce the same strain. Therefore, we could have represented this diagram in the following form:

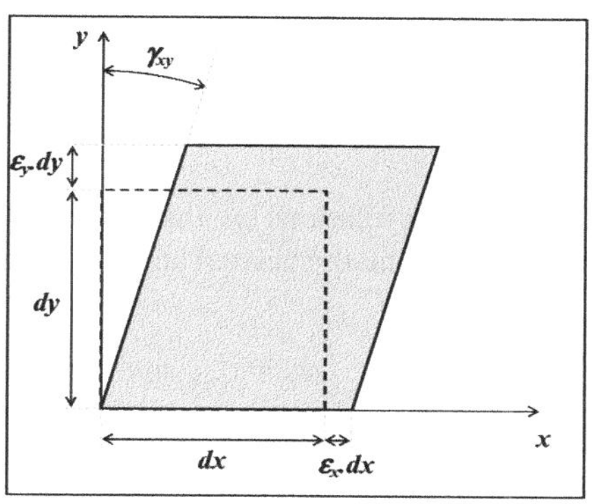

Figure 2.7. *Strain of a square*

The strain tensor enables us to completely characterize the state of a strain of a material point. In particular, it enables us to easily find the unit strain in any direction, or the angular distortion of two orthogonal vectors.

As a matter of fact, it can be shown that:

$$\begin{cases} \varepsilon(M,\underline{x}) = \varepsilon_x(M) = \underline{x}^t.\underline{\underline{\varepsilon}}(M).\underline{x} \\ \varepsilon(M,\underline{y}) = \varepsilon_y(M) = \underline{y}^t.\underline{\underline{\varepsilon}}(M).\underline{y} \\ \varepsilon(M,\underline{z}) = \varepsilon_z(M) = \underline{z}^t.\underline{\underline{\varepsilon}}(M).\underline{z} \end{cases} \quad [2.24]$$

And, whatever the vector $\underline{n}$ is, the strain in this direction is worth:

$$\varepsilon(M,\underline{n}) = \varepsilon_n(M) = \underline{n}^t.\underline{\underline{\varepsilon}}(M).\underline{n} \quad [2.25]$$

Likewise:

$$\begin{cases} \gamma(M,\underline{x},\underline{y}) = \gamma_{xy}(M) = \gamma_{yx}(M) = \underline{x}^t.\underline{\underline{\varepsilon}}(M).\underline{y} \\ \gamma(M,\underline{y},\underline{z}) = \gamma_{yz}(M) = \gamma_{yz}(M) = \underline{y}^t.\underline{\underline{\varepsilon}}(M).\underline{z} \\ \gamma(M,\underline{x},\underline{z}) = \gamma_{xz}(M) = \gamma_{zx}(M) = \underline{x}^t.\underline{\underline{\varepsilon}}(M).\underline{z} \end{cases} \quad [2.26]$$

And therefore, whatever the $\underline{n}$ and $\underline{t}$ orthogonal vectors, the distortion according to n and t is worth:

$$\gamma(M,\underline{n},\underline{t}) = \gamma_{nt}(M) = \gamma_{tn}(M) = 2.\underline{n}^t.\underline{\underline{\varepsilon}}(M).\underline{t} \quad [2.27]$$

We will notice that these two relations of the strains are very similar to the relations which enable us to determine the normal and shear stresses:

$$\begin{cases} \varepsilon_n = \varepsilon_{nn} = \underline{n}^t.\underline{\underline{\varepsilon}}.\underline{n} \\ \varepsilon_{nt} = \dfrac{\gamma_{nt}}{2} = \underline{n}^t.\underline{\underline{\varepsilon}}.\underline{t} \end{cases} \quad \text{and} \quad \begin{cases} \sigma_n = \sigma_{nn} = \underline{n}^t.\underline{\underline{\sigma}}.\underline{n} \\ \tau_{nt} = \sigma_{nt} = \underline{n}^t.\underline{\underline{\sigma}}.\underline{t} \end{cases} \quad [2.28]$$

These four relations are very practical to determine the strains, the angular distortions, the normal stresses or the shear stresses in any direction (or coordinate system), and summarize the use of stress tensors and strain tensors.

Note that ε_n and ε_{nn} are simply two different notations for speaking about the same thing, just the same as σ_n and σ_{nn} or τ_{nt} and σ_{nt}.

2.2.2. Principal strains and principal directions

As the 3D strain matrix is:

$$\underline{\underline{\varepsilon}}(M)=\begin{bmatrix}\varepsilon_x & \varepsilon_{xy} & \varepsilon_{xz}\\ \varepsilon_{xy} & \varepsilon_y & \varepsilon_{yz}\\ \varepsilon_{xz} & \varepsilon_{yz} & \varepsilon_z\end{bmatrix}_{(\underline{x},\underline{y},\underline{z})} \quad [2.29]$$

We can evidently determine this strain matrix in $(\underline{x}_1, \underline{y}_1, \underline{z}_1)$ by:

$$\underline{\underline{\varepsilon}}(M)_{B_1}=\underline{\underline{P}}^t.\underline{\underline{\varepsilon}}(M)_B.\underline{\underline{P}} \quad [2.30]$$

with $\underline{\underline{P}}$, the rotation matrix from the basis B to the basis B_1, representing the vector coordinates of B_1 expressed in the basis B.

THEOREM.– *There is a direct orthonormal coordinate system* $(\underline{x}_I, \underline{x}_{II}, \underline{x}_{III})$ *in which the strain matrix is diagonal*:

$$\underline{\underline{\varepsilon}}(M)=\begin{bmatrix}\varepsilon_I & 0 & 0\\ 0 & \varepsilon_{II} & 0\\ 0 & 0 & \varepsilon_{III}\end{bmatrix}_{(\underline{x}_I,\underline{x}_{II},\underline{x}_{III})} \quad [2.31]$$

ε_I, ε_{II}, and ε_{III} are called principal strains and $\underline{x}_I$, $\underline{x}_{II}$ and $\underline{x}_{III}$ are called principal directions , which are associated with ε_I, ε_{II} and ε_{III}, respectively.

We then clearly have:

$$\begin{cases}\underline{\underline{\varepsilon}}(M).\underline{x}_I=\varepsilon_I.\underline{x}_I\\ \underline{\underline{\varepsilon}}(M).\underline{x}_{II}=\varepsilon_{II}.\underline{x}_{II}\\ \underline{\underline{\varepsilon}}(M).\underline{x}_{III}=\varepsilon_{III}.\underline{x}_{III}\end{cases} \quad [2.32]$$

And in practical terms, to determine the principal strains, it is sufficient to write:

$$\det\left(\underline{\underline{\varepsilon}}(M)-\varepsilon_i.\underline{\underline{I}}\right)=0 \quad [2.33]$$

which gives three solutions (or two in 2D). Then, to determine the three principal directions, it is sufficient to write the three previous relations.

This can be translated by the following drawing:

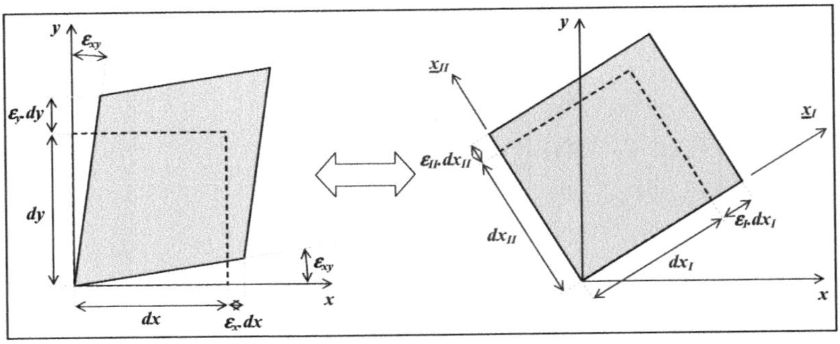

Figure 2.8. *Strain of a square and principal strains. For a color version of this figure, see www.iste.co.uk/bouvet/aeronautical.zip*

As seen by the matter on these two diagrams, the state of strain is the same.

EXAMPLES: SHEAR.–

As the strain tensor is:

$$\underline{\underline{\varepsilon}}(M)=\begin{bmatrix} 0 & \gamma/2 \\ \gamma/2 & 0 \end{bmatrix}_{(\underline{x},\underline{y})} \qquad [2.34]$$

Hence, we can show that the principal strains are $+\gamma/2$ and $-\gamma/2$ and the principal directions are oriented at +45° and −45°:

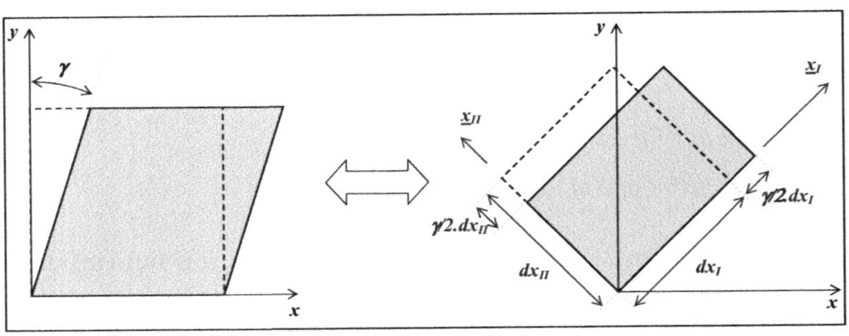

Figure 2.9. *Strain of a square and principal strains in pure shear. For a color version of this figure, see www.iste.co.uk/bouvet/aeronautical.zip*

This result can be easily understood on a physical level, as we can easily use our hands to feel that the applied force on the left diagram pulls at +45° and compresses at −45°.

This result is similar to that found for shear stresses. As a matter of fact, for isotropic materials, the shear stress simply causes shear strain.

2.2.3. *Volume expansion*

If we calculate the variation of a small rectangular section subjected to plane strain, we then find:

$$\frac{\Delta S}{S} = \frac{S' - S}{S} \approx \frac{(dx + \varepsilon_x.dx).(dy + \varepsilon_y.dy) - dx.dy}{dx.dy} \approx \varepsilon_x + \varepsilon_y \qquad [2.35]$$

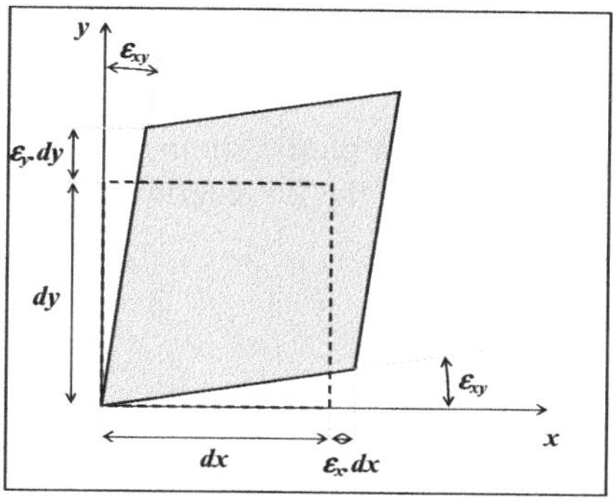

Figure 2.10. *Volume expansion*

The area variation is therefore equal to the trace of the 2D strain tensor. Evidently, this result is only true within the SPH.

The result is similar in 3D:

$$\frac{\Delta V}{V} = tr\left(\underline{\underline{\varepsilon}}\right) = \varepsilon_x + \varepsilon_y + \varepsilon_z \qquad [2.36]$$

Physically, as this volume variation is the same, regardless of the coordinate system in which the strain matrix is written, the trace must therefore be an invariant of the strain tensor, which is clearly the case.

2.2.4. *Invariants of strain tensor*

Just as for the stress tensor, the strain tensor has three elementary invariants. Classically, we use:

– the strain trace :

$$I_1 = trace\left(\underline{\underline{\varepsilon}}\right) = \varepsilon_x + \varepsilon_y + \varepsilon_z \qquad [2.37]$$

– the von Mises strain :

$$\bar{\varepsilon} = \sqrt{\frac{2}{3}.\underline{dev}\left(\underline{\underline{\varepsilon}}\right) : \underline{dev}\left(\underline{\underline{\varepsilon}}\right)} \qquad [2.38]$$

with the strain deviator, which is written as:

$$\underline{dev}\left(\underline{\underline{\varepsilon}}\right) = \underline{\underline{\varepsilon}} - \frac{1}{3}.trace\left(\underline{\underline{\varepsilon}}\right).\underline{\underline{I}} \qquad [2.39]$$

NOTE.– In order to find the equivalent plastic strain in plasticity and traction that is equal to the plastic strain in traction, the 2/3 coefficient is used instead of the 3/2 coefficient for the stress.

– the determinant:

$$I_3 = \det\left(\underline{\underline{\varepsilon}}\right) \qquad [2.40]$$

These invariants are far less common than for stresses, as the criteria are generally written on the basis of the stresses.

2.2.5. *Compatibility condition*

The strain field is derived from the displacement field:

$$\underline{\underline{\varepsilon}} = \frac{1}{2}.\left(\underline{\underline{grad}}(\underline{u}) + \underline{\underline{grad}}^t(\underline{u})\right) \qquad [2.41]$$

Therefore, in order for the strain field to be integrable, it must verify the following condition:

$$\underline{\underline{grad}}\left(\underline{div}\left(\underline{\underline{\varepsilon}}\right)\right) + \underline{\underline{grad}}^t\left(\underline{div}\left(\underline{\underline{\varepsilon}}\right)\right) - \underline{\underline{\Delta\varepsilon}} - \underline{\underline{grad}}\left(\underline{grad}\left(trace\left(\underline{\underline{\varepsilon}}\right)\right)\right) = \underline{\underline{0}} \qquad [2.42]$$

This relation, called the strain compatibility condition, is very important if we are looking to solve a problem based on the stress method (see Chapter 4). These are some of the fundamental equations that a strain field needs to verify in order to be the solution to a problem (see Chapter 4).

2.3. Strain measurement: strain gage

In order to measure the strains in a solid, we use strain gages, which are resistances stuck onto the surface. If the solid becomes strained, the resistance varies, and we can then determine the strain as seen by the material.

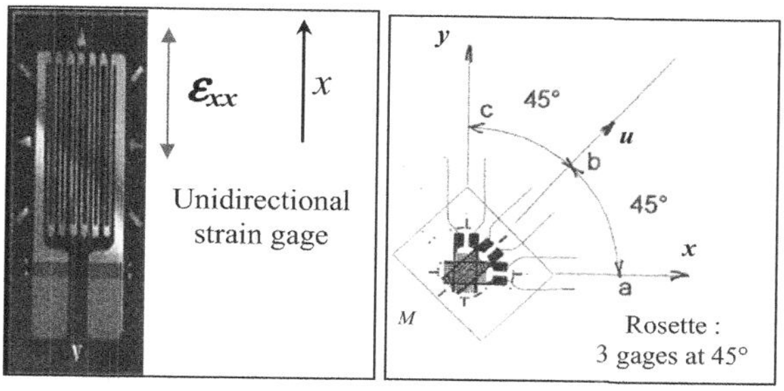

Figure 2.11. *Unidirectional gage and rosette: three gages oriented at 45°. For a color version of this figure, see www.iste.co.uk/bouvet/aeronautical.zip*

The size of a gage can vary from 1 to about 100 mm. The active parts are the bonded threads (the grid) in which the resistance is going to vary.

We can then determine the strain in the direction of the threads, and in that alone. We often use rosettes with three gages at 45°, which enable us to determine the three strains in the plane.

As a gage is stuck to a surface, it therefore only enables us to measure the plane strains:

$$\underline{\underline{\varepsilon}}(M) = \begin{bmatrix} \varepsilon_x & \varepsilon_{xy} \\ \varepsilon_{xy} & \varepsilon_y \end{bmatrix}_{(\underline{x},\underline{y})} \qquad [2.43]$$

In practical terms, knowing the strains ε_a, ε_b and ε_c of a three-gage rosette at 45° (see the previous figure), it is sufficient to write that the strain, in the directions of the three gages given by the strain tensor, is equal to the strains given by the gages:

$$\begin{cases} \underline{x}^t . \underline{\underline{\varepsilon}}(M) . \underline{x} = \varepsilon_a \\ \underline{u}^t . \underline{\underline{\varepsilon}}(M) . \underline{u} = \varepsilon_b \\ \underline{y}^t . \underline{\underline{\varepsilon}}(M) . \underline{y} = \varepsilon_c \end{cases} \qquad [2.44]$$

With:

$$\underline{u} \begin{bmatrix} 1/\sqrt{2} \\ 1/\sqrt{2} \end{bmatrix} \qquad [2.45]$$

Hence:

$$\begin{cases} \varepsilon_x = \varepsilon_a \\ \varepsilon_{xy} = \varepsilon_b - \dfrac{\varepsilon_a + \varepsilon_c}{2} \\ \varepsilon_y = \varepsilon_c \end{cases} \qquad [2.46]$$

The gages are reliable and not very expensive, so they are widely used at present.

Moreover, in practice, it is almost impossible to measure the stress, so we generally settle for estimating them based on external forces. This estimation is thus based on stress distribution hypotheses, which can be marred by error.

Strain gages are therefore the best way to measure the strain of a structure. If we know the behavior law (see the next chapter) of the material, the stresses can be determined based on the strain. Due to malapropism, strain gage is incidentally often called stress gage!

3

Behavior Law

3.1. A few definitions

– Behavior law: this is the relation that links the stress and strain. Evidently, it depends on the material and the external conditions (temperature, hygrometry, etc.).

– Homogeneous: a material is said to be homogeneous if it behaves the same behavior at each point. In practice, it is never evidently neither completely true nor false. For example, a folded metal sheet will have different behavior in the area of folding as the material will have a different history.

– Isotropic: a material is said to be isotropic if it behaves in the same way in all directions. In practice, it is never evidently neither completely true nor false. For example, a laminated metal sheet will have different behavior in its plane and in accordance with the normal direction as the lamination will have hardened (see section 3.3) this direction more substantially than the other two directions.

In this course, we will (almost) restrict ourselves to isotropic homogeneous materials.

3.2. Tension test

This is a sample with an applied force in tension:

We measure the tension stress σ_x based on the force F by:

$$\sigma_x = \frac{F}{S} \quad [3.1]$$

and we measure the strains ε_x and ε_y based on two strain gages.

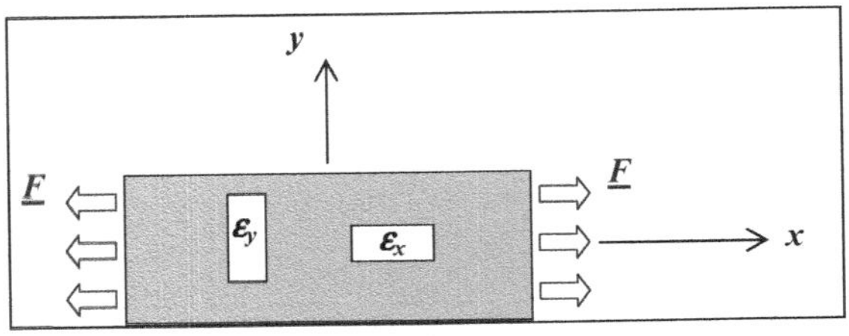

Figure 3.1. *Tension test. For a color version of this figure, see www.iste.co.uk/bouvet/aeronautical.zip*

3.2.1. *Brittle materials*

Brittle materials are: glass, ceramics, carbon, composites based on glass or carbon fibers, steels with a high elastic limit, etc.

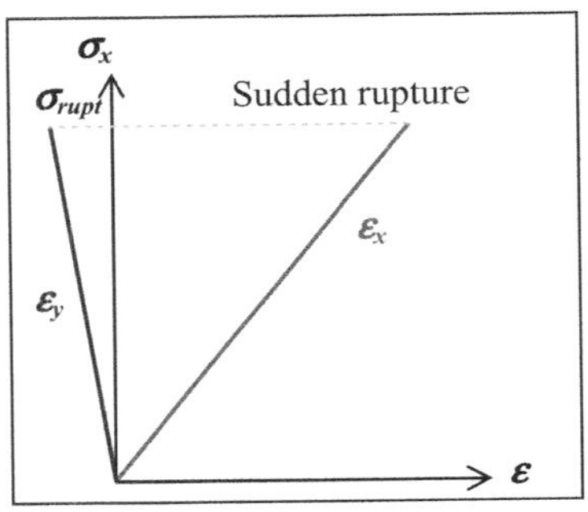

Figure 3.2. *Tensile test of a brittle material. For a color version of this figure, see www.iste.co.uk/bouvet/aeronautical.zip*

Quasi-linear curves are obtained until there is a sudden rupture. This behavior is linear elastic; elastic as in it all returns to its point of departure. As a matter of fact, we can find materials with nonlinear elastic behavior (rubber, etc.).

So, we adopt a linear elastic behavior law model:

$$\begin{cases} \sigma_x = E.\varepsilon_x \\ \varepsilon_y = -\nu.\varepsilon_x \end{cases} \quad [3.2]$$

with E being Young's modulus (in MPa). For example, 210 GPa for the steel, 60 GPa for the glass, 200 GPa for the carbon and ν for Poisson's ratio (without a unit), which is typically equal to about 0.3 for "standard" materials.

Be careful, you must bear in mind that this is just a behavior model; in other words, we observe that the curves are quasi-linear, so we model them with a refined law. However, this is just a model and in reality, behavior is always much more complex.

3.2.2. *Ductile materials*

Ductile materials are mainly metals: steel, copper, aluminum, etc.

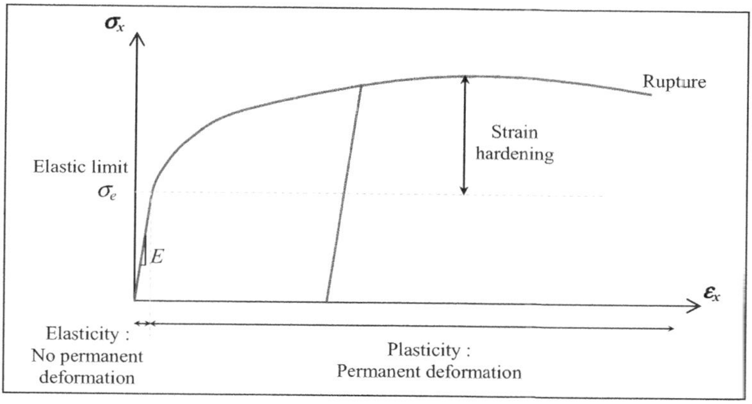

Figure 3.3. *Tension test of a ductile material. For a color version of this figure, see www.iste.co.uk/bouvet/aeronautical.zip*

We obtain a curve with a linear elastic beginning, then the rest of the curve is substantially nonlinear with a plasticity; plasticity in the sense where when discharging, it does not return to the point of departure and a permanent deformation is maintained under zero stress.

We can therefore adopt a linear behavior law if, and only if, the stress remains lower than the elastic limit:

– if $\sigma_x < \sigma_e$ $\quad \begin{cases} \sigma_x = E.\varepsilon_x \\ \varepsilon_y = -\nu.\varepsilon_x \end{cases}$

– if not, we define a behavior law with plasticity.

3.2.3. *Particular cases*

Be careful, you must bear in mind that these behavior laws are but models which pale in comparison to reality, which is always much more complex when looked at closely.

And even when one does not look too closely, there are certain behaviors which are more complex. For example, rubbers are elastic but substantially nonlinear; polymers are substantially viscous (they are a kind of intermediate between solids and liquids).

3.3. Shear test

This is a sample with an applied force in shear:

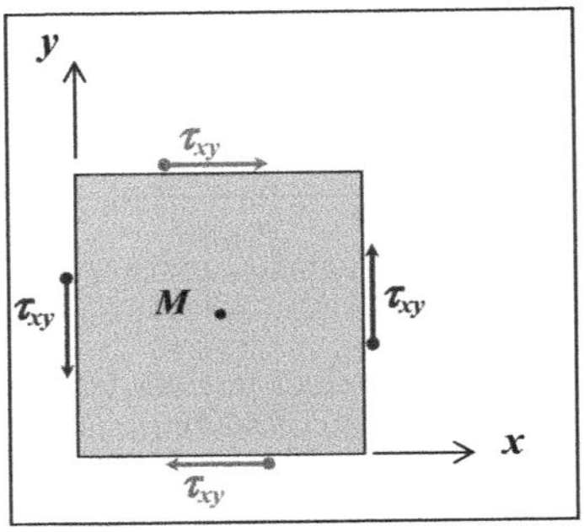

Figure 3.4. *Shear. For a color version of this figure, see www.iste.co.uk/bouvet/aeronautical.zip*

Evidently, the problem is much more complicated than applying a force to a material in tension. In practice, we use a cylinder with an applied force of torsion on a thin wall. It can therefore be shown that the applied force is mainly in shear stress ($\tau_{z\theta}$) and that it is homogeneous. We can therefore make an estimate based on the measurement of the torsion torque. The shear strain ($\gamma_{z\theta}$) therefore remains to be determined, which is based on the strain gages stuck to the external wall (we can show that the gages at +45° and −45° fit the bill).

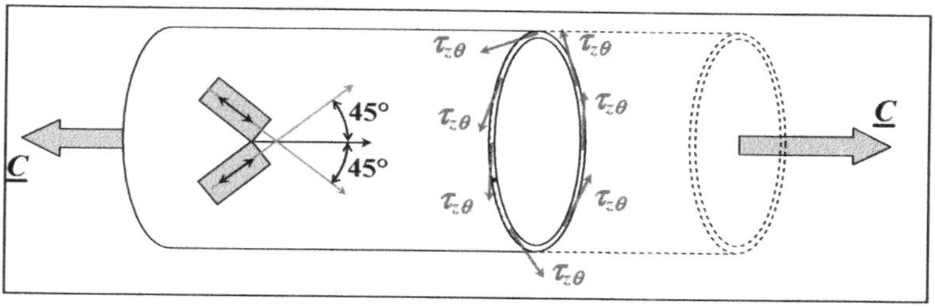

Figure 3.5. *Torsion test. For a color version of this figure, see www.iste.co.uk/bouvet/aeronautical.zip*

3.3.1. *Brittle materials*

We obtain quasi-linear curves until a sudden rupture. This behavior is linear elastic. We therefore adopt a linear elastic behavior law:

$$\tau = G.\gamma \qquad [3.3]$$

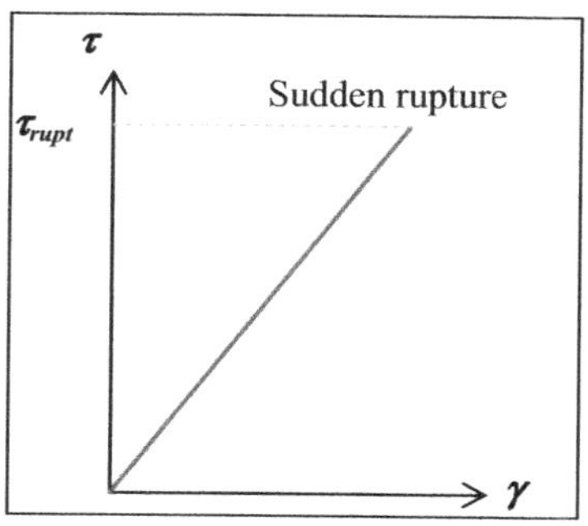

Figure 3.6. *Brittle material shear test. For a color version of this figure, see www.iste.co.uk/bouvet/aeronautical.zip*

with G being the shear modulus (in MPa). And for an isotropic linear elastic material, we can show that:

$$G = \frac{E}{2.(1+\nu)} \qquad [3.4]$$

3.3.2. *Ductile materials*

Just as in the previous instance, we obtain a curve with an elastic beginning, and then the rest is plastic.

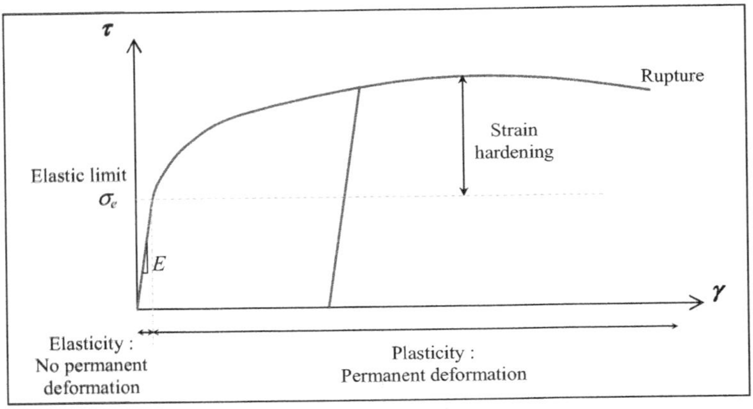

Figure 3.7. *Shear test of a ductile material. For a color version of this figure, see www.iste.co.uk/bouvet/aeronautical.zip*

We can therefore adopt a linear behavior law if, and only if, the stress remains lower than the elastic limit:

– If $\tau < \tau_e$ $\quad\quad \tau = G.\gamma$

– If not, we define a behavior law with plasticity.

And we experimentally take note that τ_e is approximately equal to half of σ_e.

3.4. General rule

3.4.1. *Linear elasticity*

Let us begin with 2D. If we give ourselves:

$$\underline{\underline{\sigma}}(M) = \begin{bmatrix} \sigma_{xx} & \tau_{xy} \\ \tau_{xy} & \sigma_{yy} \end{bmatrix}_{(\underline{x},\underline{y})} \quad [3.5]$$

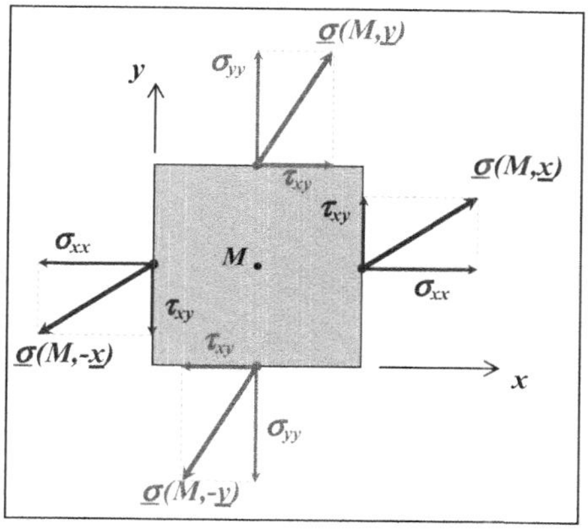

Figure 3.8. *Stress vectors on the faces of a square. For a color version of this figure, see www.iste.co.uk/bouvet/aeronautical.zip*

So, in the linear part (before any possible plasticity), we adopt an isotropic linear elastic behavior:

$$\begin{cases} \varepsilon_x = \dfrac{1}{E}.\left(\sigma_x - \nu.\sigma_y\right) \\ \varepsilon_y = \dfrac{1}{E}.\left(\sigma_y - \nu.\sigma_x\right) \\ \varepsilon_{xy} = \dfrac{\tau_{xy}}{2.G} = \dfrac{(1+\nu).\tau_{xy}}{E} \end{cases} \qquad [3.6]$$

Or by reversing:

$$\begin{cases} \sigma_x = \dfrac{E}{1-\nu^2}.\left(\varepsilon_x + \nu.\varepsilon_y\right) \\ \sigma_y = \dfrac{E}{1-\nu^2}.\left(\varepsilon_y + \nu.\varepsilon_x\right) \\ \tau_{xy} = 2.G.\varepsilon_{xy} = \dfrac{E}{1+\nu}.\varepsilon_{xy} \end{cases} \qquad [3.7]$$

And in 3D:

$$\begin{cases} \varepsilon_x = \dfrac{1}{E}.\left(\sigma_x - \nu.\left(\sigma_y + \sigma_z\right)\right) \\ \varepsilon_y = \dfrac{1}{E}.\left(\sigma_y - \nu.\left(\sigma_x + \sigma_z\right)\right) \\ \varepsilon_z = \dfrac{1}{E}.\left(\sigma_z - \nu.\left(\sigma_x + \sigma_y\right)\right) \\ \varepsilon_{xy} = \dfrac{\tau_{xy}}{2.G} = \dfrac{(1+\nu).\tau_{xy}}{E.} \\ \varepsilon_{yz} = \dfrac{\tau_{yz}}{2.G} = \dfrac{(1+\nu).\tau_{yz}}{E} \\ \varepsilon_{xz} = \dfrac{\tau_{xz}}{2.G} = \dfrac{(1+\nu).\tau_{xz}}{E} \end{cases} \quad [3.8]$$

Or by reversing:

$$\begin{cases} \sigma_x = 2.\mu.\varepsilon_x + \lambda.\left(\varepsilon_x + \varepsilon_y + \varepsilon_z\right) \\ \sigma_y = 2.\mu.\varepsilon_y + \lambda.\left(\varepsilon_x + \varepsilon_y + \varepsilon_z\right) \\ \sigma_z = 2.\mu.\varepsilon_z + \lambda.\left(\varepsilon_x + \varepsilon_y + \varepsilon_z\right) \\ \tau_{xy} = 2.G.\varepsilon_{xy} = \dfrac{E}{(1+\nu)}.\varepsilon_{xy} \\ \tau_{yz} = 2.G.\varepsilon_{yz} = \dfrac{E}{(1+\nu)}.\varepsilon_{yz} \\ \tau_{xz} = 2.G.\varepsilon_{xz} = \dfrac{E}{(1+\nu)}.\varepsilon_{xz} \end{cases} \quad [3.9]$$

With λ and μ the Lamé parameters:

$$\begin{cases} \lambda = \dfrac{E.\nu}{(1+\nu).(1-2.\nu)} \\ \mu = \dfrac{E}{2.(1+\nu)} = G \end{cases} \quad [3.10]$$

Or even more simply written in tensorial form:

$$\underline{\underline{\sigma}} = 2.\mu.\underline{\underline{\varepsilon}} + \lambda.trace\left(\underline{\underline{\varepsilon}}\right).\underline{\underline{I}} \qquad [3.11]$$

This relation is fundamental, as it translates the relationship between stress and strain (in other words, it allows us to see whether the material is rigid or not). It is one of the fundamental equations that must verify the stress/strain fields in order to be the solution to the problem (see Chapter 4).

If we reverse this relation, we then find:

$$\underline{\underline{\varepsilon}} = \frac{1+\nu}{E}\underline{\underline{\sigma}} - \frac{\nu}{E}.trace\left(\underline{\underline{\sigma}}\right).\underline{\underline{I}} \qquad [3.12]$$

And by making some calculations, we can find the Lamé parameters by identifying these two relations.

EXAMPLES: VOLUME VARIATION IN TENSION.–

It can be shown that in tension:

$$\underline{\underline{\sigma}} = \begin{bmatrix} \sigma_0 & 0 & 0 \\ 0 & 0 & 0 \\ 0 & 0 & 0 \end{bmatrix}_{(\underline{x},\underline{y},\underline{z})} \qquad [3.13]$$

And therefore:

$$\underline{\underline{\varepsilon}} = \begin{bmatrix} \frac{\sigma_0}{E} & 0 & 0 \\ 0 & \frac{-\nu.\sigma_0}{E} & 0 \\ 0 & 0 & \frac{-\nu.\sigma_0}{E} \end{bmatrix}_{(\underline{x},\underline{y},\underline{z})} \qquad [3.14]$$

The variation of volume is therefore:

$$\frac{\Delta V}{V} = trace\left(\underline{\underline{\varepsilon}}\right) = \frac{(1-2.\nu).\sigma_0}{E} \qquad [3.15]$$

And as ν is generally more or less equal to 0.3, we observe that the volume increases in a tension test.

EXAMPLES: FREE EXTERNAL SURFACE.–

On an external surface, with a normal $\underline{z}$, which is free from stress, we have:

$$\underline{\sigma}(M,\underline{n}_{ext}) = \underline{\sigma}(M,\underline{z}) = \underline{\underline{\sigma}}(M).\underline{z} = \underline{0} \quad [3.16]$$

The stress tensor is therefore in this form:

$$\underline{\underline{\sigma}} = \begin{bmatrix} \sigma_{xx} & \tau_{xy} & 0 \\ \tau_{xy} & \sigma_{yy} & 0 \\ 0 & 0 & 0 \end{bmatrix}_{(\underline{x},\underline{y},\underline{z})} \quad [3.17]$$

This is called a plane stress. To put it plainly, there is no stress vector outside the plane that is exerted on this external surface.

However, we can show that for an isotropic linear elastic material, strains are therefore in the form below:

$$\underline{\underline{\varepsilon}} = \begin{bmatrix} \varepsilon_{xx} & \varepsilon_{xy} & 0 \\ \varepsilon_{xy} & \varepsilon_{yy} & 0 \\ 0 & 0 & \varepsilon_{zz} \end{bmatrix}_{(\underline{x},\underline{y},\underline{z})} \quad [3.18]$$

With:

$$\varepsilon_{zz} = -\frac{\nu}{E}.(\sigma_{xx} + \sigma_{yy}) \quad [3.19]$$

In other words, we can never have plane stress and plane strain at the same time.

In practice, in order to simplify problems, where possible, we try to position themselves in plane stress or plane strain. Evidently, these are only the hypotheses that remain to be verified or which have been verified based on previous experiments.

For example, if we study a thin plate, it can be shown that the stresses are planes on the two external surfaces, so at first approximation, they will be so in practice as well. And in this case, the strain will not then be plane.

On the other hand, if we study a thick plate, we can show that the plane strain hypothesis will be better suited to the problem. And in this case, the stress will not then be plane.

3.5. Anisotropic materials: example of a composite

So that you do not being to think that all materials are isotropic, we will now study a particular material, a composite made up of 50% (in volume) carbon fibers and 50% epoxy resin. This is classically used in aeronautics; it is T300/914. It is in the form of a 0.1 to 0.2 mm thick fabric which we will later cut, and then crape, in order to obtain the desired thickness.

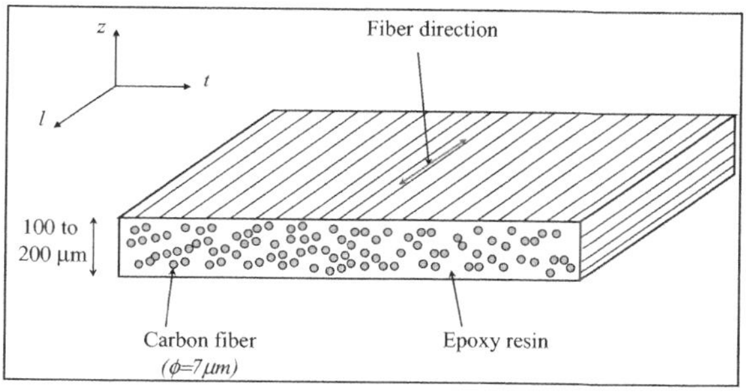

Figure 3.9. *Unidirectional composite carbon/epoxy. For a color version of this figure, see www.iste.co.uk/bouvet/aeronautical.zip*

3.5.1. *Elasticity*

Let us put the laminate plane in 2D $(\underline{l}, \underline{t})$; l for longitudinal and t for transversal directions. If we perform a tension test with the fibers in the θ-direction, we evidently note that the elasticity modulus depends on θ:

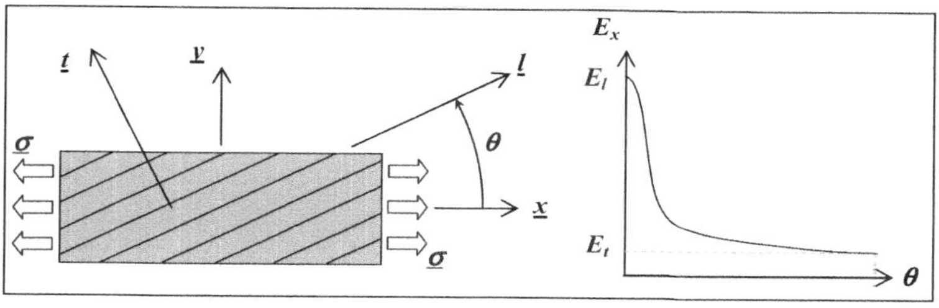

Figure 3.10. *Young's modulus of a UD composite versus the tension direction. For a color version of this figure, see www.iste.co.uk/bouvet/aeronautical.zip*

We can therefore simply write the behavior law in the coordinate $(\underline{l}, \underline{t})$:

$$\begin{cases} \varepsilon_l = \dfrac{\sigma_l}{E_l} - \dfrac{v_{lt}.\sigma_t}{E_l} \\ \varepsilon_t = \dfrac{\sigma_t}{E_t} - \dfrac{v_{lt}.\sigma_l}{E_l} \\ \varepsilon_{lt} = \dfrac{\tau_{lt}}{2.G_{lt}} \end{cases} \qquad [3.20]$$

Evidently, this is a little more complicated in $(\underline{x}, \underline{y})$. And for the T300/914 considered, we have:

$$\begin{cases} E_l = 130\ GPa \\ E_t = 10\ GPa \\ \upsilon_{lt} = 0.35 \\ G_{lt} = 4.5\ GPa \end{cases} \qquad [3.21]$$

3.6. Thermoelasticity

In the instance where the temperature of a free stress structure varies, we evidently observe thermal expansion phenomena. At first approximation, this expansion is translated by the strain that is proportional to this temperature variation:

$$\underline{\underline{\varepsilon_{th}}} = \alpha.\Delta T.\underline{\underline{I}} \qquad [3.22]$$

where α is the coefficient of thermal expansion (in K^{-1}), and the identity tensor signifies that the strain is homogeneous in every direction (if this strain is not constrained).

In order to factor in this thermally originated strain in elasticity, it is sufficient to add it to the elastic strain:

$$\underline{\underline{\varepsilon}} = \underline{\underline{\varepsilon_e}} + \underline{\underline{\varepsilon_{th}}} = \underline{\underline{\varepsilon_e}} + \alpha.\Delta T.\underline{\underline{I}} \qquad [3.23]$$

And the previous behavior relation remains valid if it is written with the elastic strain, and no longer with the total strain:

$$\underline{\underline{\sigma}} = 2.\mu.\underline{\underline{\varepsilon_e}} + \lambda.trace\left(\underline{\underline{\varepsilon_e}}\right).\underline{\underline{I}} = 2.\mu.\underline{\underline{\varepsilon}} + \lambda.trace\left(\underline{\underline{\varepsilon}}\right).\underline{\underline{I}} - \frac{E}{1-2.\upsilon}.\alpha.\Delta T.\underline{\underline{I}} \qquad [3.24]$$

And by reversing it:

$$\underline{\underline{\varepsilon_e}} = \frac{1+\nu}{E}\underline{\underline{\sigma}} - \frac{\nu}{E}.trace\left(\underline{\underline{\sigma}}\right).\underline{\underline{I}} \qquad [3.25]$$

Or even:

$$\underline{\underline{\varepsilon}} = \underline{\underline{\varepsilon_e}} + \underline{\underline{\varepsilon_{th}}} = \frac{1+\nu}{E}\underline{\underline{\sigma}} - \frac{\nu}{E}.trace\left(\underline{\underline{\sigma}}\right).\underline{\underline{I}} + \alpha.\Delta T.\underline{\underline{I}} \qquad [3.26]$$

EXAMPLES: CONSTRAINED OR FREE EXPANSION.–

If we consider a material that is free from stress and subjected to a temperature variation, which is placed in 2D:

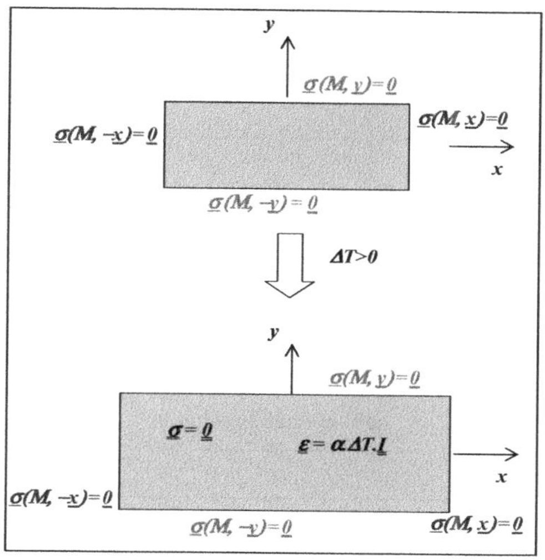

Figure 3.11. *Thermal expansion with free stress. For a color version of this figure, see www.iste.co.uk/bouvet/aeronautical.zip*

We can therefore show that the stress is homogeneous and null:

$$\underline{\underline{\sigma}} = \begin{bmatrix} 0 & 0 \\ 0 & 0 \end{bmatrix}_{(\underline{x},\underline{y})} \qquad [3.27]$$

And that the structure is therefore subjected to a homogeneous strain field:

$$\begin{cases} \underline{\underline{\varepsilon}} = \underline{\underline{\varepsilon_{th}}} = \begin{bmatrix} \alpha.\Delta T & 0 \\ 0 & \alpha.\Delta T \end{bmatrix}_{(\underline{x},\underline{y})} \\ \underline{\underline{\varepsilon_{el}}} = \begin{bmatrix} 0 & 0 \\ 0 & 0 \end{bmatrix}_{(\underline{x},\underline{y})} \end{cases} \qquad [3.28]$$

However, if we consider that the displacement is constrained:

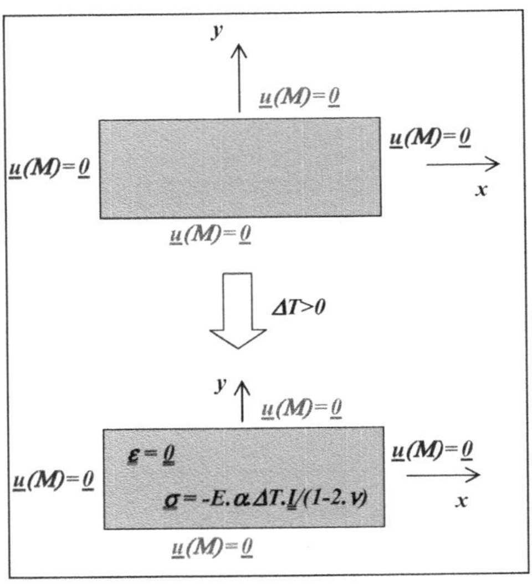

Figure 3.12. *Thermal expansion with constrained strain. For a color version of this figure, see www.iste.co.uk/bouvet/aeronautical.zip*

We can therefore show that the total strain is homogeneous and null, and that the elastic and thermal strains will balance each other out:

$$\begin{cases} \underline{\underline{\varepsilon}} = \begin{bmatrix} 0 & 0 \\ 0 & 0 \end{bmatrix}_{(\underline{x},\underline{y})} \\ \underline{\underline{\varepsilon_{th}}} = \begin{bmatrix} \alpha.\Delta T & 0 \\ 0 & \alpha.\Delta T \end{bmatrix}_{(\underline{x},\underline{y})} \\ \underline{\underline{\varepsilon_{el}}} = \begin{bmatrix} -\alpha.\Delta T & 0 \\ 0 & -\alpha.\Delta T \end{bmatrix}_{(\underline{x},\underline{y})} \end{cases} \qquad [3.29]$$

The structure is therefore subjected to a homogeneous bi-axial compression stress field:

$$\underline{\underline{\sigma}} = \begin{bmatrix} -\dfrac{E}{1-2.\upsilon}.\alpha.\Delta T & 0 \\ 0 & -\dfrac{E}{1-2.\upsilon}.\alpha.\Delta T \end{bmatrix}_{(\underline{x},\underline{y})} \qquad [3.30]$$

Any readers who are interested may find details of these calculations in the exercise chapter (Beam thermal expansion: mono-material beam).

4

Resolution Methods

4.1. Assessment

The study of deformable solids consists of determining the following at each point of the structure:

$$\begin{cases} \underline{\underline{\sigma}}(M) \\ \underline{\underline{\varepsilon}}(M) \\ \underline{u}(M) \end{cases} \qquad [4.1]$$

There are 15 unknown functions.

There are three relations that are:

– the relation between the displacement and the strain;

– the equilibrium equation;

– the material's behavior law:

$$\begin{cases} \underline{\underline{\varepsilon}} = \frac{1}{2}.\left(\underline{\underline{grad}}(\underline{u}) + \underline{\underline{grad}}^t(\underline{u})\right) \\ \underline{div}\left(\underline{\underline{\sigma}}\right) + \underline{f}_v = \underline{0} \\ \underline{\underline{\sigma}} = 2.\mu.\underline{\underline{\varepsilon}} + \lambda.trace\left(\underline{\underline{\varepsilon}}\right).\underline{\underline{I}} \end{cases} \qquad [4.2]$$

These three relations have completely different natures:

– The first stems directly from the definition of the strain tensor. In this form, it only varies with SPH, but it is completely possible to write it without too much difficulty in other instances as well. This is an exact relation!

– The second one is a principle and it is therefore true all of the time. We can evidently adapt it to the dynamic.

– The third is a model and it depends on the material. It is given here in its most simple form possible, which is of a linear elastic homogeneous isotropic material; of course, it is generally much more complex than this. The problem of how to write these behavior laws is broadly unresolved, which is the subject of much research at present.

There are therefore 15 scalar equations. This problem of multiple equations can therefore be resolved by adding the boundary conditions in order to determine the integration constants.

The displacement boundary conditions where the displacements are imposed can be written as:

$$\underline{u}(M) = \underline{u}_{imp} \qquad [4.3]$$

And the stress boundary conditions where the external forces are imposed can be written as:

$$\underline{\underline{\sigma}}(M, \underline{n}_{ext}) = \underline{F}_{ext} \qquad [4.4]$$

Evidently, this problem is far from simple to solve and the problems which can be solved manually are exceptional cases! All the same, we are lucky to know that the solution to our problem does exist and that it is unique, so if we manage to find one, then we can be sure that it is correct!

Finally, we must not lose sight of the objective of studying deformable solids, which is to size and design structures. Once the stress, strain and displacement fields have been determined, we can then apply the sizing criterion to the structure (rupture criterion, non-permanent deformation criterion, etc.) and to determine whether or not the current structure meets the specifications. Incidentally, the sizing criterion used needs to be pertinent, which is not necessarily a given.

From there:

– either the structure does not meet the specifications and therefore the shape or material needs to be modified, then the previous calculations need to be redone with the new structure;

– or the structure meets the specifications and perhaps it can be built, or perhaps we deem that it resists external forces too well and that it is possible to make it lighter (aeronautics). We can modify its shape or its material, and then redo the previous calculations with the new structure.

Obviously, this is an iterative approach.

4.2. Displacement method

This method consists of postulating a form for the displacement field. This field obviously needs to verify the displacement boundary conditions.

We then verify that this field verifies the equilibrium equation which, in the instance of linear elastic isotropic behavior, can be written via the intermediary of Navier's equation:

$$\mu.\underline{\Delta}(\underline{u})+(\lambda+\mu).\underline{grad}(div(\underline{u}))+\underline{f}_v=\underline{0} \quad [4.5]$$

We then determine the strain, then the stress, and finally, we verify the stress boundary conditions.

Once again, as the solution to our problem exists and is unique (i.e. if the boundary limits are properly laid out), then if we find a solution, it must be the correct one! All that remains now is to have some good ideas for postulating a pertinent displacement field.

4.3. Stress method

This method consists of postulating a form for the stress field. This field obviously needs to verify the stress boundary conditions.

We then verify that the stress field verifies the equilibrium equation:

$$\underline{div}(\underline{\underline{\sigma}})+\underline{f}_v=\underline{0} \quad [4.6]$$

We then determine the strain via the intermediary of the behavior law. It now therefore remains to integrate this field in order to determine the displacement, but in practice, in order for this field to be integrable, it must verify the compatibility equation, either written in strain as:

$$\underline{\underline{grad}}\left(\underline{div}\left(\underline{\underline{\varepsilon}}\right)\right)+\underline{\underline{grad}}^{t}\left(\underline{div}\left(\underline{\underline{\varepsilon}}\right)\right)-\underline{\underline{\Delta\varepsilon}}-\underline{\underline{grad}}\left(\underline{grad}\left(trace\left(\underline{\underline{\varepsilon}}\right)\right)\right)=\underline{\underline{0}} \qquad [4.7]$$

or written in stress, which gives the case of linear elastic isotropic behavior:

$$\underline{\underline{\Delta\sigma}}+\frac{1}{1+\nu}.\underline{\underline{grad}}\left(\underline{grad}\left(trace\left(\underline{\underline{\sigma}}\right)\right)\right)=\frac{-\nu}{1-\nu}.div\left(\underline{f}_{v}\right).\underline{\underline{I}}-\left(\underline{\underline{grad}}^{t}\left(\underline{f}_{v}\right)+\underline{\underline{grad}}\left(\underline{f}_{v}\right)\right) \qquad [4.8]$$

It can be shown that these two scripts are equivalent.

The strain can also be integrated to determine the displacement field, and it therefore remains to verify the displacement boundary conditions.

Once again, as the solution to our problem exists and is unique (i.e. if the boundary limits are properly laid out), then if we find a solution, it must be the correct one! All that remains now is to have some good ideas for postulating a pertinent stress field.

4.4. Finite element method

As we are unable to resolve the problem in an exact manner for the majority of cases, we have to resort to solving them in an approximate manner. This is, for example, the objective of the finite element method, which shall be introduced throughout the following chapter. Currently, this method is the most commonly used one by far and it enables us to size almost all types of structure.

5

Work-energy Theorem: Principle of Finite Element Method

5.1. Work-energy theorem

5.1.1. *Hypotheses*

This is a solid S:

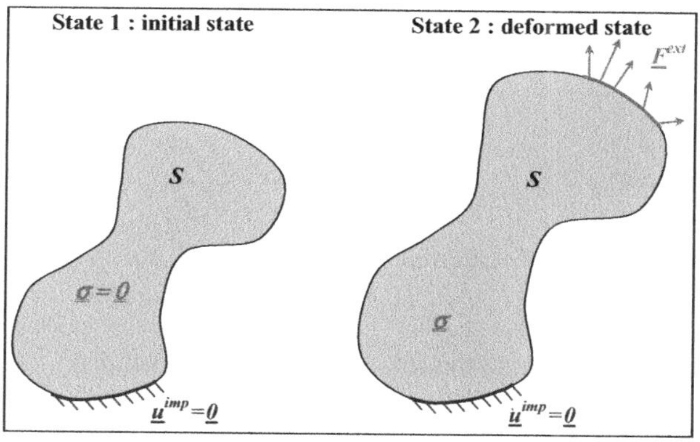

Figure 5.1. *Definition of the initial and deformed states. For a color version of this figure, see www.iste.co.uk/bouvet/aeronautical.zip*

State 1 is what is called the initial state (or stress-free state), which is free from stress and state 2 is the deformed state, which is under the effect of external forces.

We adopt the following hypotheses:

– the gravity is neglected (otherwise state 1 would not be stress-free);

– the behavior is assumed to be linear elastic;

– both the external heat exchanges (heat input, etc.) and internal heat exchanges (no friction, etc.) are neglected;

– the external forces are applied infinitely slowly between 1 and 2 (hence, there is no inertia);

– the displacement and the strain are small (SPH) and it does not affect the directions of the external forces.

5.1.2. *Strain energy*

The kinetic energy theorem between states 1 and 2 is written as:

$$W_{ext}^{1\to 2} + W_{\text{int}}^{1\to 2} = \Delta E_c^{1\to 2} + \Delta E_p^{1\to 2} = 0 \qquad [5.1]$$

In other words, the sum of the work of external and internal forces on S between the states 1 and 2 is null and equal to the sum of kinetic energy variations and its potential energy between states 1 and 2. In short, the energy is conserved!

Hence:

$$W_{ext}^{1\to 2} = -W_{\text{int}}^{1\to 2} \qquad [5.2]$$

Therefore, the work of external forces is equal to the stored energy in S $(-W_{int}^{1\to 2})$.

This relation is equivalent to the first law of thermodynamics:

$$du + dq = \delta w \qquad [5.3]$$

Integrated between states 1 and 2, neglecting the heat exchanges:

$$\Delta u^{1\to 2} = W_{ext}^{1\to 2} \qquad [5.4]$$

So, once again, the internal energy variation ($\Delta u^{1 \to 2}$) in the solid between 1 and 2 is equal to the work of the external forces and, in particular, does not depend:

– on the behavior law of the material;

– on the way in which the external forces are applied.

In particular, in the mechanics of deformable solids, the internal energy only depends on the strain state.

If we also choose $u_1 = 0$, then:

$$W_{ext} = U_2 = E_d \qquad [5.5]$$

And E_d is called the strain energy (in J).

5.1.3. *Work of external forces*

If we consider n external forces applied at points $P_1,\ldots, P_n$, with the displacement $\underline{u}_1,\ldots, \underline{u}_n$, then the work of external forces is written as:

$$W_{ext} = \sum_{i=1}^{n} W_{ext}{}^{i} = \sum_{i=1}^{n} \int_1^2 \underline{F}_i(t).d\underline{u}_i(t) \qquad [5.6]$$

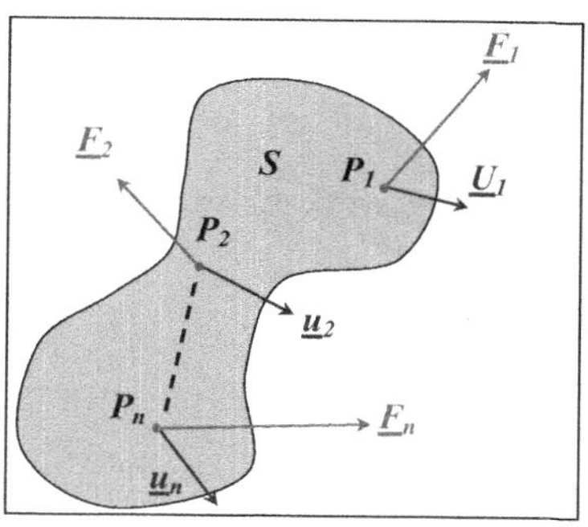

Figure 5.2. *External forces. For a color version of this figure, see www.iste.co.uk/bouvet/aeronautical.zip*

Of course, in reality, each $\underline{u}_p(t)$ continuously moves from $\underline{0}$ to $\underline{u}_p$ and each $\underline{F}_i(t)$ moves continuously from $\underline{0}$ to $\underline{F}_i$. Moreover, the behavior of the structure is linear:

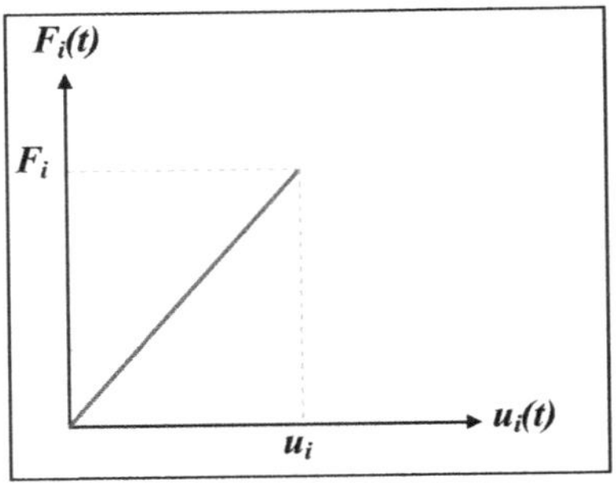

Figure 5.3. *Linearity of external forces. For a color version of this figure, see www.iste.co.uk/bouvet/aeronautical.zip*

So:

$$\int_1^2 \underline{F}_i(t).d\underline{u}_i(t) = \frac{1}{2}\underline{F}_i.\underline{u}_i \tag{5.7}$$

And therefore:

$$W_{ext} = \frac{1}{2}\sum_{i=1}^{n} \underline{F}_i.\underline{u}_i \tag{5.8}$$

5.1.4. *Strain energy*

By using the analogy with the work of external forces, we define the strain energy by:

$$E_d = \frac{1}{2}.\iiint_V \underline{\underline{\sigma}} : \underline{\underline{\varepsilon}}.dV \tag{5.9}$$

Or, in its developed form:

$$E_d = \frac{1}{2}.\iiint_V \left(\sigma_{xx}.\varepsilon_x + \sigma_{yy}.\varepsilon_y + \sigma_{zz}.\varepsilon_z + 2.\tau_{xy}.\varepsilon_{xy} + 2.\tau_{yz}.\varepsilon_{yz} + 2.\tau_{xz}.\varepsilon_{xz}\right).dV \tag{5.10}$$

In this relation, the role of the force is played by the stress and the role of the displacement is played by the strain.

The triple integral on volume impresses, but simply signifies that all of the points of the solid are added up.

EXAMPLE: TENSION.–

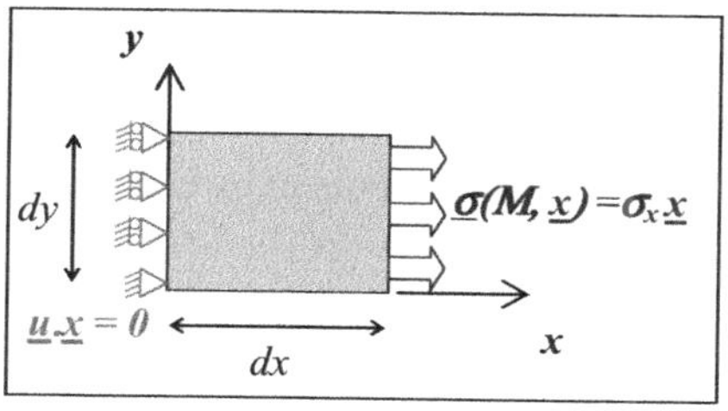

Figure 5.4. *Tension test. For a color version of this figure, see www.iste.co.uk/bouvet/aeronautical.zip*

In order to imagine this, we will apply it to the tension example. Evidently, we have:

$$\underline{\underline{\sigma}} = \begin{bmatrix} \sigma_x & 0 & 0 \\ 0 & 0 & 0 \\ 0 & 0 & 0 \end{bmatrix} \qquad [5.11]$$

– Calculation of the work of external forces:

On the faces with normal vectors $\pm\underline{y}$ and $\pm\underline{z}$, the external forces are null, so the work is also null. On the face with normal vector $\underline{x}$, the force is in the x-direction and the displacement is null according to this direction so the work is null. Hence:

$$W_{ext} = \frac{1}{2}.F_x.u_x \qquad [5.12]$$

where F_x is the force in the x-direction and u_x the displacement in the x-direction of the face with normal vector $\underline{x}$.

And the strain energy is worth:

$$E_d = \frac{1}{2}.\sigma_x.\varepsilon_x.dx.dy.dz \qquad [5.13]$$

Hence:

$$\begin{cases} F_x = \sigma_x.dy.dz \\ u_x = \varepsilon_x.dx \end{cases} \quad [5.14]$$

In conclusion, it works (the work of external forces is definitely equal to the strain energy), and the expression of strain energy can even be demonstrated in this way by working on the stress tensor, component by component.

5.1.5. *Energy minimization: Ritz method*

We are looking to approximately solve the following problem:

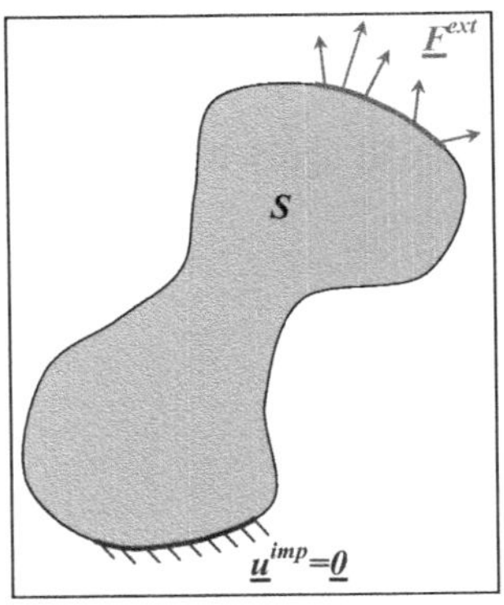

Figure 5.5. *Setting the problem. For a color version of this figure, see www.iste.co.uk/bouvet/aeronautical.zip*

So, at each point M, we look for the stress, strain and displacement fields.

As it is generally not possible to solve the problem exactly, we shall solve it approximately. Therefore, we look for the displacement in a pre-defined form:

$$\underline{u}(M) = \underline{f}(M, c_1, c_2, ..., c_n) \quad [5.15]$$

where the function $\underline{f}$ is a known function that depends on the parameters c_1, c_2,…,c_n which are unknown.

We have therefore reduced the search down to n parameters, rather than a search for three unknown functions; the three displacements (in 3D). Evidently, the problem is simpler, but the drawback is that the form of the function f is assumed *a priori.*

If this function is properly chosen and there are a set of parameters representing the exact solution of the problem, then we have:

$$E_d = W_{ext} \qquad [5.16]$$

If the exact solution is not in this form, then the difference:

$$E_d - W_{ext} \qquad [5.17]$$

will be minimum for the best choice of parameters c_1, c_2,...,c_n. In fact, we can show that this difference must necessarily be positive and null for the exact solution.

The relations:

$$\begin{cases} \dfrac{\partial(E_d - W_e)}{\partial c_1} = 0 \\ \cdots\cdots \\ \dfrac{\partial(E_d - W_e)}{\partial c_n} = 0 \end{cases} \qquad [5.18]$$

enable us to determine these parameters.

So, the energy minimization enables choosing the best function among all the functions that you have given yourself. The advantage is that you will definitely have a solution, but the disadvantage is that if the set of functions that you have chosen is far from the exact solution, then this minimization will give you the one which is least bad: which is not necessarily a correct one!

5.2. Finite element method

5.2.1. *General principle of finite element method*

The principle of the finite element method is to discretize the real problem in simple domains on which the displacement field is a simple function (called the basis function or the interpolation function) of displacements in nodes (edges of the domains).

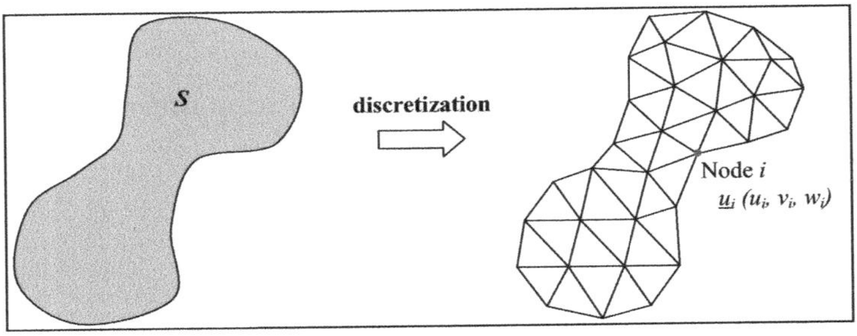

Figure 5.6. *Discretization of the structure in finite element method. For a color version of this figure, see www.iste.co.uk/bouvet/aeronautical.zip*

So, if we have n nodes:

$$\underline{u}(M) = \begin{vmatrix} u(M) \\ v(M) \\ w(M) \end{vmatrix} = \begin{bmatrix} \varphi_{11} & \varphi_{12} & \varphi_{13} & \cdots & \varphi_{1(3.n)} \\ \varphi_{21} & & & & \\ \varphi_{31} & & & & \varphi_{3(3.n)} \end{bmatrix} \cdot \begin{bmatrix} u_1 \\ v_1 \\ w_1 \\ \cdots \\ \cdots \\ u_n \\ v_n \\ w_n \end{bmatrix} = \underline{\underline{\varphi}}(M).\underline{U} \qquad [5.19]$$

where $\underline{\underline{\varphi}}(M)$ represents the matrix of the basis functions, which depends on M, but it is totally known, and $\underline{U}$ represents the displacement vector at nodes, and therefore corresponds to the parameters to be determined, i.e. c_1, c_2,…,c_n from the previous section.

Then in order to determine the strain field, we derive the displacement field, so the basis functions $\underline{\underline{\varphi}}(M)$ can be expressed as:

$$\underline{\varepsilon}(M) = \underline{\underline{\delta}}(M).\underline{U} \qquad [5.20]$$

where the notation $\underline{\varepsilon}(M)$ corresponds to the strain components placed in the column:

$$\underline{\varepsilon}(M) = \begin{bmatrix} \varepsilon_x \\ \varepsilon_y \\ \varepsilon_z \\ \gamma_{xy} \\ \gamma_{yz} \\ \gamma_{xz} \end{bmatrix} \qquad [5.21]$$

NOTE.– *we note γ_{xy} rather than ε_{xy}, which will enable us to obtain the strain energy by making a scalar product with the stress vector.*

This strain notation in the form of a vector is evidently mathematically false; it is just a notation to simplify the scripts. In particular, the classic properties of vectors (for example, the rules on rotation) do not apply to the strain which is definitely a matrix!

And $\underline{\underline{\delta}}(M)$ corresponds to the derivative matrix of the basis functions $\underline{\underline{\phi}}(M)$; evidently as the vector $\underline{U}$ is constant, its derivative is null.

Then we determine the stress:

$$\underline{\sigma}(M) = \underline{\underline{L}}.\underline{\varepsilon}(M) = \underline{\underline{L}}.\underline{\underline{\delta}}(M).\underline{U} \qquad [5.22]$$

As with the strain, the stress vector is:

$$\underline{\sigma}(M) = \begin{bmatrix} \sigma_x \\ \sigma_y \\ \sigma_z \\ \tau_{xy} \\ \tau_{yz} \\ \tau_{xz} \end{bmatrix} \qquad [5.23]$$

And $\underline{\underline{L}}$ is the rigidity matrix:

$$\underline{\underline{L}} = \begin{bmatrix} 2.\mu+\lambda & \lambda & \lambda & 0 & 0 & 0 \\ \lambda & 2.\mu+\lambda & \lambda & 0 & 0 & 0 \\ \lambda & \lambda & 2.\mu+\lambda & 0 & 0 & 0 \\ 0 & 0 & 0 & \mu & 0 & 0 \\ 0 & 0 & 0 & 0 & \mu & 0 \\ 0 & 0 & 0 & 0 & 0 & \mu \end{bmatrix} \qquad [5.24]$$

which enables us to obtain the behavior law introduced in the previous chapter.

Then we calculate the strain energy:

$$E_d = \frac{1}{2}.\iiint_V \underline{\sigma}^t(M).\underline{\varepsilon}(M).dV \qquad [5.25]$$

Hence:

$$E_d = \frac{1}{2}.\underline{U}^t.\underline{\underline{K}}.\underline{U} \qquad [5.26]$$

with $\underline{\underline{K}}$ being the rigidity matrix of the whole structure:

$$\underline{\underline{K}} = \iiint_V \underline{\underline{\delta}}^t(M).\underline{\underline{L}}.\underline{\underline{\delta}}(M).dV \qquad [5.27]$$

This matrix obviously does not depend on the point M and represents the entire behavior of the structure. We can show that it is a square and diagonal matrix whose size depends on the number of nodes chosen.

The work of external forces remains to be calculated:

$$W_{ext} = \frac{1}{2}.\underline{F}^t.\underline{U} \qquad [5.28]$$

where the vector $\underline{F}$ is the vector of the external forces on the nodes where the forces are applied, and 0 elsewhere.

The minimization of the energy therefore gives:

$$\frac{\partial(E_d - W_e)}{\partial \underline{U}} = \underline{0} \qquad [5.29]$$

Hence:

$$\underline{\underline{K}}.\underline{U} - \frac{1}{2}.\underline{F} - \frac{1}{2}.\frac{\partial \underline{F}}{\partial \underline{U}}.\underline{U} = \underline{0} \qquad [5.30]$$

and

$$\underline{\underline{K}}.\underline{U} = \underline{F} \qquad [5.31]$$

Knowing that $\underline{U}$ is the unknown:

$$\underline{U} = \underline{\underline{K}}^{-1}.\underline{F} \qquad [5.32]$$

In conclusion, to solve a problem with the finite element method, you must start by modeling your boundary conditions in force and in displacement. As a matter of fact, in practice, we do not know them precisely and the engineer's job is to choose the boundary conditions for his or her model, which represent the real boundary conditions as best as possible. This stage is essential and often very delicate!

We then discretize the structure in elementary sub-domains. Once again, this approach is not simple and it is laden with consequences as it largely conditions the quality of the final results. In general, the greater the number of nodes, the more precise the solution will be, but this is not always true. However, this poses the problem of the calculation time, which increases with the number of nodes.

We then select a behavior law, and once again, we need to make the wisest approximation possible.

Finally, what remains is to perform the calculation *per se*, of which the largest part consists of inverting the structure's rigidity matrix.

5.2.2. *Example of the three-node triangular element*

Let us put this in 2D and consider a triangular finite element with three nodes: 1, 2 and 3. There are six unknowns in the problem, namely the displacements to the three nodes: u_1, v_1, u_2, v_2, u_3, v_3:

$$\underline{U} = \begin{bmatrix} u_1 \\ v_1 \\ u_2 \\ v_2 \\ u_3 \\ v_3 \end{bmatrix} \qquad [5.33]$$

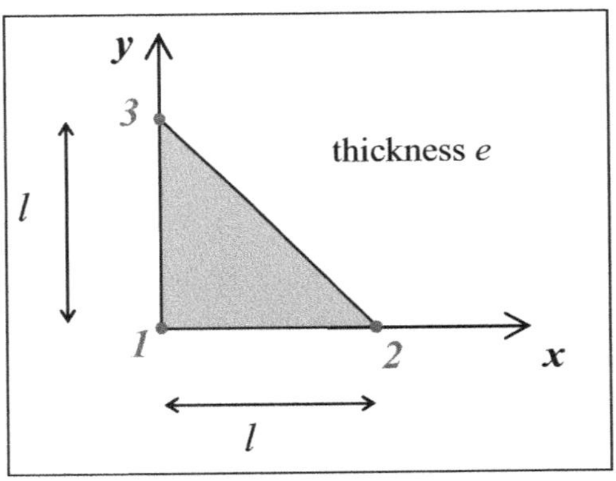

Figure 5.7. *Triangular finite element. For a color version of this figure, see www.iste.co.uk/bouvet/aeronautical.zip*

We choose a linear displacement field which enables us to find the value of the displacements at the nodes:

$$\underline{u}(M) = \begin{vmatrix} u(x,y) \\ v(x,y) \end{vmatrix} = \begin{vmatrix} u_1 + (u_2 - u_1).\dfrac{x}{l} + (u_3 - u_1).\dfrac{y}{l} \\ v_1 + (v_2 - v_1).\dfrac{x}{l} + (v_3 - v_1).\dfrac{y}{l} \end{vmatrix} \qquad [5.34]$$

Or even:

$$\underline{u}(M)=\underline{\underline{\varphi}}(M).\underline{U}=\begin{bmatrix}1-\frac{x}{l}-\frac{y}{l} & 0 & \frac{x}{l} & 0 & \frac{y}{l} & 0\\ 0 & 1-\frac{x}{l}-\frac{y}{l} & 0 & \frac{x}{l} & 0 & \frac{y}{l}\end{bmatrix}.\begin{bmatrix}u_1\\ v_1\\ u_2\\ v_2\\ u_3\\ v_3\end{bmatrix} \quad [5.35]$$

We can then determine the strain field:

$$\underline{\varepsilon}(M)=\begin{bmatrix}\varepsilon_x\\ \varepsilon_y\\ \gamma_{xy}\end{bmatrix}=\begin{bmatrix}\frac{\partial u}{\partial x}\\ \frac{\partial v}{\partial y}\\ \frac{\partial u}{\partial y}+\frac{\partial v}{\partial x}\end{bmatrix}=\begin{bmatrix}\frac{u_2-u_1}{l}\\ \frac{v_3-v_1}{l}\\ \frac{u_3-u_1+v_2-v_1}{l}\end{bmatrix} \quad [5.36]$$

Or even:

$$\underline{\varepsilon}(M)=\underline{\underline{\delta}}(M).\underline{U}=\frac{1}{l}.\begin{bmatrix}-1 & 0 & 1 & 0 & 0 & 0\\ 0 & -1 & 0 & 0 & 0 & 1\\ -1 & -1 & 0 & 1 & 1 & 0\end{bmatrix}.\begin{bmatrix}u_1\\ v_1\\ u_2\\ v_2\\ u_3\\ v_3\end{bmatrix} \quad [5.37]$$

Here the matrix $\underline{\underline{\delta}}(M)$ is constant, but it is obviously a particular case.

We then determine the stress and then the rigidity matrix of the structure (I am going to skip a few lines of calculations):

$$\underline{\underline{K}} = \frac{E.e}{2.\left(1-\nu^2\right)}.\begin{bmatrix} \frac{3-\nu}{2} & \frac{\nu+1}{2} & -1 & \frac{\nu-1}{2} & \frac{\nu-1}{2} & -\nu \\ \frac{\nu+1}{2} & \frac{3-\nu}{2} & -\nu & \frac{\nu-1}{2} & \frac{\nu-1}{2} & -1 \\ -1 & -\nu & 1 & 0 & 0 & \nu \\ \frac{\nu-1}{2} & \frac{\nu-1}{2} & 0 & \frac{1-\nu}{2} & \frac{1-\nu}{2} & 0 \\ \frac{\nu-1}{2} & \frac{\nu-1}{2} & 0 & \frac{1-\nu}{2} & \frac{1-\nu}{2} & 0 \\ -\nu & -1 & \nu & 0 & 0 & 1 \end{bmatrix} \quad [5.38]$$

This is a symmetrical matrix that needs to be inverted in order to determine $\underline{U}$.

If, for example, we look at the following problem:

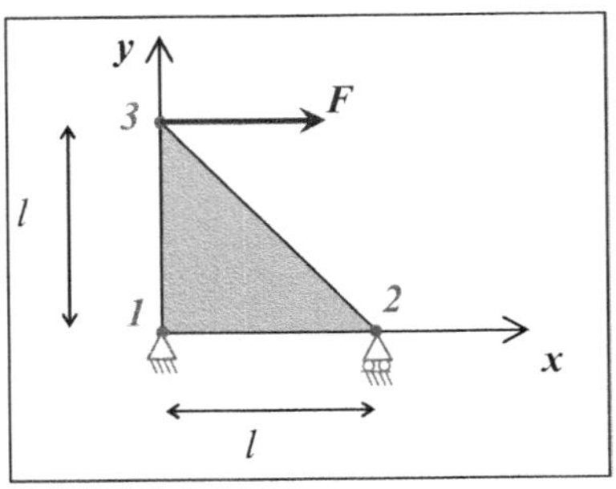

Figure 5.8. *Triangle subjected to a force. For a color version of this figure, see www.iste.co.uk/bouvet/aeronautical.zip*

Here, we therefore have:

$$\begin{cases} u_1 = 0 \\ v_1 = 0 \\ v_2 = 0 \end{cases} \quad [5.39]$$

We incidentally notice that these boundary conditions enable us to block all rigid body displacement field, and that the solution will therefore be unique.

We have:

$$\underline{U} = \begin{vmatrix} u_2 \\ u_3 \\ v_3 \end{vmatrix} \tag{5.40}$$

And

$$\underline{F} = \begin{vmatrix} 0 \\ F \\ 0 \end{vmatrix} \tag{5.41}$$

Then:

$$\underline{\underline{K}} = \frac{E.e}{2.\left(1-\nu^2\right)}.\begin{bmatrix} 1 & 0 & \nu \\ 0 & \frac{1-\nu}{2} & 0 \\ \nu & 0 & 1 \end{bmatrix} \tag{5.42}$$

Hence:

$$\underline{U} = \underline{\underline{K}}^{-1}.\underline{F} = \frac{1+\nu}{E}.\begin{bmatrix} 0 \\ F \\ 0 \end{bmatrix} \tag{5.43}$$

Hence, the following displacement field:

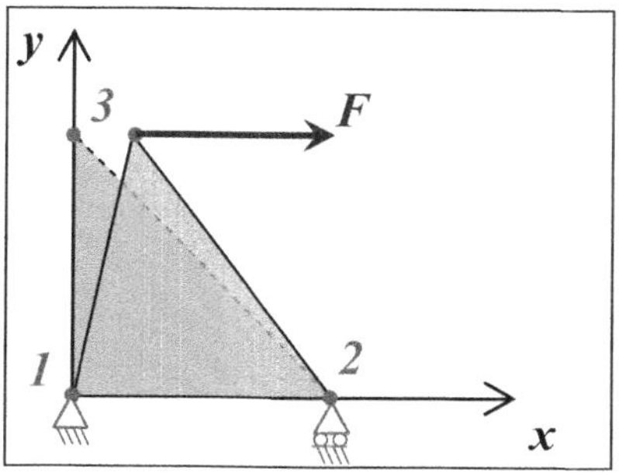

Figure 5.9. *Displacement field of a triangle subjected to a force. For a color version of this figure, see www.iste.co.uk/bouvet/aeronautical.zip*

And:

$$\underline{\varepsilon}(M) = \begin{bmatrix} \varepsilon_x \\ \varepsilon_y \\ \gamma_{xy} \end{bmatrix} = \begin{bmatrix} 0 \\ 0 \\ \dfrac{(1+\nu).F}{E.e.l} \end{bmatrix} \qquad [5.44]$$

$$\underline{\sigma}(M) = \begin{bmatrix} \sigma_x \\ \sigma_y \\ \tau_{xy} \end{bmatrix} = \begin{bmatrix} 0 \\ 0 \\ \dfrac{2.F}{e.l} \end{bmatrix} \qquad [5.45]$$

In short, only the point 3 is displaced in the x-direction, and we have the creation of a shear strain, as well as that of a shear stress. The value of this stress is equal to the force F divided by (e.l/2), or the section of the plate at mid-height.

If you are interested, you can re-do this very simple problem on a finite element code, such as Catia. You can, for example, follow the guide given in the next section.

If we now pose the problem of an applied force on a plate in this way, we will obviously find a more complex displacement field. This is due to the fact that the discretization adopted here imposes a linear displacement field. If, on the other hand, we choose a finer grid, we will then obtain a more realistic displacement field:

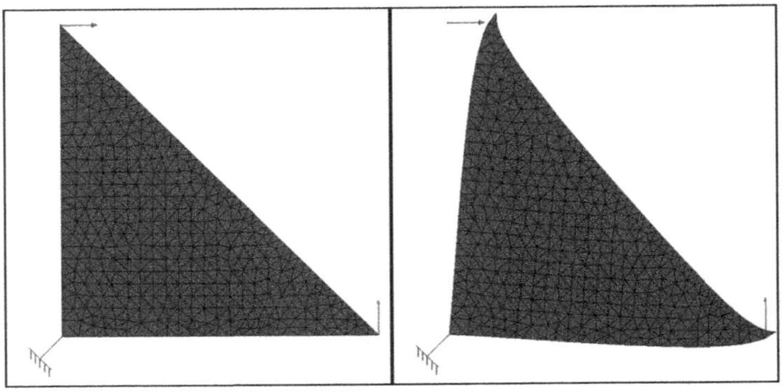

Figure 5.10. *Triangle subjected to a force and displacement field obtained by FE calculation. For a color version of this figure, see www.iste.co.uk/bouvet/aeronautical.zip*

And the stress fields are also more realistic:

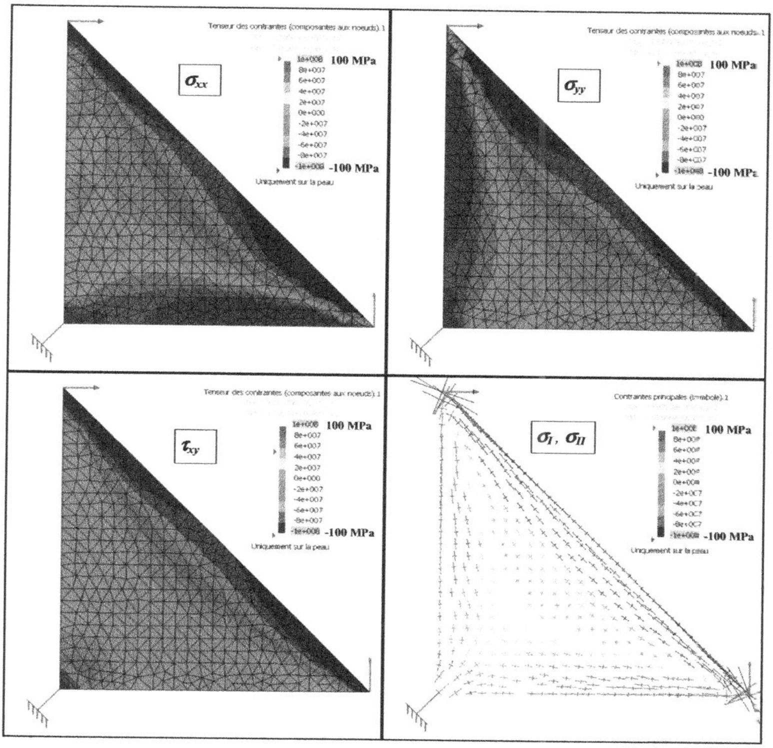

Figure 5.11. *Stress field determined by FE calculation in a triangle subjected to a force. For a color version of this figure, see www.iste.co.uk/bouvet/aeronautical.zip*

Be careful; despite the fact that the external force and the boundary conditions are applied in points, the stress will tend to infinity as the grid is refined! In short, this result is available only far away from the boundary conditions; this is Saint-Venant's principle.

Physically, the principal stress diagram is very instructive. In the central part, we mainly observe shearing, or equal principal stresses of opposite signs at +45° and –45°. We also observe a significant compression force on the line going from node 2 to node 3 and significant tension on the lines going from nodes 1 to 2 and 1 to 3.

In conclusion, in order to design this structure, it would be worthwhile for us to put a plate in the central part so as to withstand the shear and beams on the three sides of the triangle so as to withstand the tension and compression forces.

5.3. Application: triangle with plate finite element using Catia

The problem is as follows:

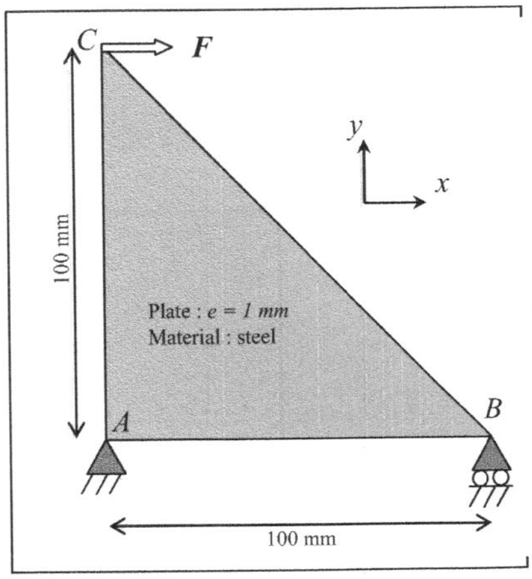

Figure 5.12. *Problem of a triangle subjected to a force*

Start by opening a "Part Design", open a "Sketch" in the (x, y) plane and then create the triangle.

Go to the "Wireframe and Surface Design" and carry out a "Fill" of the sketch. Then, if Catia does not do it automatically, hide ("Hide/Show") the sketch.

Then apply the steel material ("Apply Material") by selecting the "Part Body" of your piece.

Go to the "Generative Structural Analysis" module then choose the "Static Analysis".

Start by applying a surface grid with "Octree Triangle Mesher" (hidden behind "Octree Tetrahedron Mesher"). You can then change the size of the grids by double-clicking on "Octree Triangle Mesh.1" on the "Nodes and Elements" branch of the tree. As there is only one element, choose 200 mm, and if you want to find the example of this lesson, choose the "Linear" elements (the "Parabolic" elements add an intermediary node to the center of each side and therefore enable us to account for a linear stress variation on the element). These (parabolic) elements have to be used absolutely when performing your future finite element calculations with Catia.

Select "2D Property", then select your grid and fill in the thickness. By default, the material is the one you have attributed to your "Part Body".

Then apply a "Clamp" at point A. To block the displacement according to Y of point B, choose "User-defined Restraint" then deselect all the settings except that corresponding to Y.

Then apply a force ("Distributed Force") at C of 1,000 N.

You can start the calculation ("Compute") by selecting the "All" option if it is not already done, then observe the displacement ("Deformation").

Catia automatically chooses the scale factor of the deformation. To choose 1 and therefore obtain the displacement in the correct size, go to "Amplification Magnitude". To obtain the displacement vector of each node, go to "Displacement" (Be careful: Catia automatically changes the scale factor of the deformation back. To block this, tick "Set as default for future created images").

Select "Principal Stress" (hidden behind "Displacement"), then double-click in the tree on the "Principal Stresses". Click on "More" then choose "Center of element" in Position. You now have the principal stresses that we have calculated throughout. You can also change the display scale of the stress by double-clicking on the colored scale and changing the maximum and minimum.

To observe the other stresses, after having double clicked in the tree on "Principal Stresses", choose "Average Iso", then "Tensor Component". Then click on "More", and in "Component" you can choose the three components of the 2D tensor (If you have done a 3D calculation, you will have six components). Ensure that you properly choose the global axis system if it is not already done. To do this, click on the three small dots in "Axis system", and then choose the "Global" axis system in "Type".

You can redo the same analysis by changing the grid.

6

Sizing Criteria of an Aeronautical Structure

6.1. Introduction

Now that we have seen how to determine stress and strain in a structure (at least in the simplest cases – for real cases, an approximate solution method, such as that of finite element, should be used), we shall now look at sizing criteria. As a reminder (see Introduction), the aim of the mechanics of deformable solids is to study the internal state of the material (notion of stress) and the way in which it becomes strained (notion of strain) in order to determine if the piece or the mechanical system studied attains rupture, or if it does not become too strained. In short, once the stress and strain in a piece are known, then the criteria (typically, these are the norms of the stress and/or the strain tensor) need to be applied in order to determine whether or not the piece breaks! It is these criteria (or to be more exact, some criteria, as there are hundreds of them) which shall be the subject of this chapter.

However, we must bear in mind that these criteria are not absolute truths, but are simply models which attempt to translate reality. This reality is always much more complex when looked at closer up. In short, a model is considered to be correct as long as you have not been able to show that it is false, or more specifically, as long as all of the experimental tests have not proven it to be defective!

We must also bear in mind that throughout the life of a structure, it shall be subjected to different loads (and not just to one, such as in academic exercises), called “load case”. The engineer’s job is therefore to guarantee that the structure will withstand all of the load cases without breaking, and/or without becoming too strained.

The aeronautics case is even more complex, as we distinguish limit loads (LL) from ultimate loads (UL). The sizing of aeronautical structures is indeed very complex (far too complex to cover in this book), but at first estimate, we can succinctly summarize it by saying that the structure must statically withstand the LL without incurring any damage (and/or plasticity for metallic materials) and withstand the UL without breaking in a catastrophic manner [BOU 16]. To put it more precisely, the limit loads (LL) are defined as loads that the structure will, on average, see once in their lifetime, or for aeronautics, with a probability of occurrence of about 10^{-5} per flight hour (a plane flies at about 10^5 h). As for ultimate loads (UL), they are defined as very improbable loads, or for aeronautics, with a probability of occurrence of about 10^{-9} per flight hour (ACJ 25-1309). In practice, there is a ratio k, which is generally comprised between 1.1 and 1.5 among both:

$$CE = k.CL \tag{6.1}$$

The philosophy of sizing an aeronautical structure could therefore be summarized as:

– no damage or permanent deformation is permitted in service, namely for realistic loads or even those less than or equal to the LL;

– the structure must remain whole for a test load, namely for very improbable loads or even those less than or equal to the UL.

All of the damage problems should also be added to this. As well as withstanding the LL and UL, you must in fact guarantee that the structure can undergo damage (object drop, impact of runway debris, of birds, lightning, etc.) without causing a catastrophic rupture; this is the notion of damage tolerance (which, once again, exceeds the framework of this book).

Nevertheless, regardless of how complex the sizing of a structure may be, this always goes to show that the structure (possibly with damage, cracks or missing parts) undergoes such a load without breaking or straining in an exaggerated manner. The sizing criteria should therefore be applied to the stresses and/or the strains (or on the energy restitution rates) to demonstrate that the structure can withstand the load. These criteria must obviously be verified everywhere in the structure and for all the load cases. The engineer's job then consists of modifying the design of the structure if this is not the case, or if the criteria are too far from the allowable limits (in short if the structure is too oversized and therefore too heavy).

We will now move on to review the different sizing criteria by specifying which type of material is applied each time. However, before doing that, we are going to see that determining a criterion that is adapted to each material is a long and complex process.

6.2. Experimental determination of a sizing criterion

The problem is as follows: does the state of a given stress lead to a material becoming ruptured and/or becoming irreversibly damaged and/or to plasticity appearing? It is therefore a problem with six dimensions, just like the stress tensor:

$$\underline{\underline{\sigma}} = \begin{bmatrix} \sigma_{xx} & \tau_{xy} & \tau_{xz} \\ \tau_{xy} & \sigma_{yy} & \tau_{yz} \\ \tau_{xz} & \tau_{yz} & \sigma_{zz} \end{bmatrix}_{(\underline{x},\underline{y},\underline{z})} \qquad [6.2]$$

It can be translated by an equation of the following type:

$$f\left(\underline{\underline{\sigma}}\right) \leq \sigma_0 \qquad [6.3]$$

where σ_0 is a stress limit and $f(\underline{\underline{\sigma}})$ is a norm of the stress tensor which may possibly involve parameters, depending on the material.

In the instance that the material is isotropic, we can therefore reduce it to a 3D problem by placing it within the principal stress coordinate system. This also means that if the material is orthotropic, this will not be the case!

First, we will study an isotropic material and put the problem in 2D to simplify it. The stress tensor can therefore be written thus in the principal coordinate system:

$$\underline{\underline{\sigma}}(M) = \begin{bmatrix} \sigma_I & 0 \\ 0 & \sigma_{II} \end{bmatrix}_{(\underline{x}_I,\underline{x}_{II})} \qquad [6.4]$$

To apply this stress state to our material, we obviously need to load it progressively, using a testing machine, for example. We will therefore start by choosing a load type, then progressively load the material with the same type of load. This is proportional loading:

$$\underline{\underline{\sigma}}(t) = \begin{bmatrix} \sigma(t) & 0 \\ 0 & a.\sigma(t) \end{bmatrix}_{(\underline{x}_I,\underline{x}_{II})} \qquad [6.5]$$

The problem can then be reduced to determining the value of $\sigma(t)$ leading to the rupture (or at the start of the damage/plasticity), and for all the values of $(a \in]-\infty,+\infty[)$. Graphically, this goes back to finding the admissible stress on the affine straight lines:

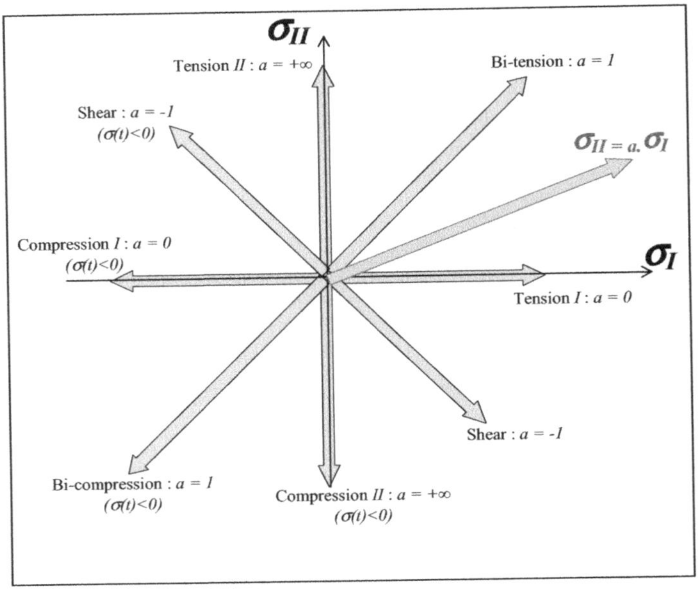

Figure 6.1. *Proportional loading. For a color version of this figure, see www.iste.co.uk/bouvet/aeronautical.zip*

In order to perform these loads, we can use a bi-tension test, for example:

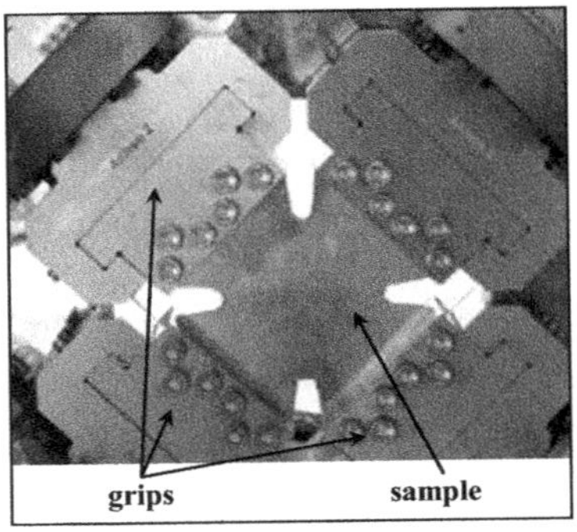

Figure 6.2. *Bi-axial testing machine (source: http://www.directindustry.fr/prod/zwick). For a color version of this figure, see www.iste.co.uk/bouvet/aeronautical.zip*

Of course, these tests are not simple to perform. In particular, we need to ensure that the test enables us to obtain a homogeneous stress area and to measure the stress in this area (either based on the external forces, as is the tension case, or based on extensometric gages, but the behavior law of the material must be known).

We must also verify the repeatability of the tests by performing the same test several times. In fact, two experimental tests never give exactly the same result! For example, for composite material, each test is performed at least five times in order to determine the rupture criteria with the associated rupture probability.

Once the tests have been performed with a sufficient amount of different values, we can then trace the criterion and compare it with the theoretical criteria in order to determine which criterion is best suited to the material studied.

Finally, the problem is even more complex in 3D (the tests can be very complex) as well as for orthotropic materials, as we need to imagine tests with a shear applied force on the material.

6.3. Normal stress or principal stress criterion: brittle material

The simplest criterion is that of the maximum normal stress. It maintains that the rupture of the material has been reached if the maximum normal stress (for all facets or all possible normal vectors $\underline{n}$) reaches a limit stress, determined using a tension test.

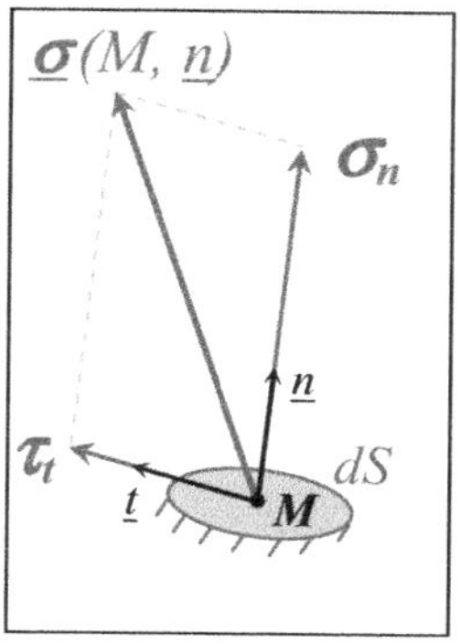

Figure 6.3. *Normal and shear stress. For a color version of this figure, see www.iste.co.uk/bouvet/aeronautical.zip*

We can also show (I will leave you to do the demonstration yourselves) that this maximum normal stress is the maximum of the three principal stresses. The rupture will therefore be developed in a plane that is perpendicular to the maximum normal stress, meaning perpendicular to the maximum principal stress.

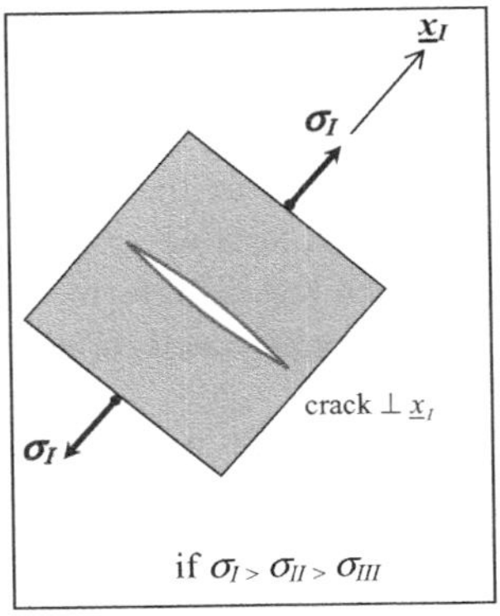

Figure 6.4. *Rupture of a brittle material. For a color version of this figure, see www.iste.co.uk/bouvet/aeronautical.zip*

Thus, we consider that we do not have a rupture if:

$$Max(\sigma_I, \sigma_{II}, \sigma_{III}) < \sigma_0 \quad [6.6]$$

with σ_0 being the rupture limit stress or the start of the damage (we can show that this criterion makes no sense for materials with plasticity, meaning for all ductile materials).

This criterion is relevant for the tension loading, meaning for the positive principal stresses; but it is often less relevant for compression loading. We do, however, use it for compression loads, but we then distinguish the tension rupture from the compression rupture. We then consider that we do not have a rupture if:

$$\begin{cases} Max(\sigma_I, \sigma_{II}, \sigma_{III}) < \sigma_{trac} \\ Min(\sigma_I, \sigma_{II}, \sigma_{III}) > \sigma_{comp} \end{cases} \quad [6.7]$$

with σ_{trac} and σ_{comp} expressing the limit tension stress and limit compression stress (considered to be negative) respectively. For example, for an epoxy resin (which may be considered a brittle material at first approximation), we have 50 MPa of tension limit and –200 MPa of compression limit. The criterion obtained in the plane (σ_I, σ_{II}) is therefore a rectangle:

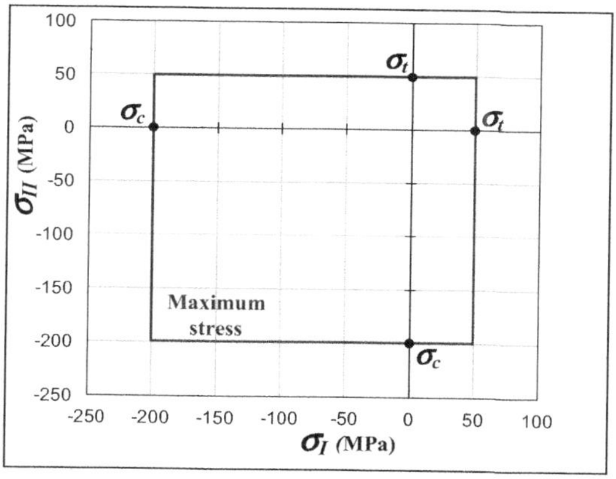

Figure 6.5. *Rupture criterion of a brittle material. For a color version of this figure, see www.iste.co.uk/bouvet/aeronautical.zip*

At first approximation, this criterion is adapted to the brittle materials, but it presents various limitations.

The main limitation is its lack of coupling between directions. In short, the limit on rupture in one direction does not depend on stresses in other directions. In order to make this type of criteria a default, we can, for example, perform a shear test by applying a force to the material with two equal stresses that have opposite signs:

$$\underline{\underline{\sigma}} = \begin{bmatrix} \sigma & 0 & 0 \\ 0 & -\sigma & 0 \\ 0 & 0 & 0 \end{bmatrix} \qquad [6.8]$$

Obviously, this load is equivalent to a pure shear in a coordinate system turned at 45° around the z-axis.

In general, we also show that the rupture stress is lower than the rupture tension stress. This comes from the fact that, in order to provoke the rupture, the compression stress in the y-direction "adds itself" to the tension stress in the x-direction. This instance will be detailed throughout the part on criteria of shear stress with friction.

Note that this criterion makes no sense for materials with plasticity, meaning for ductile materials.

As a matter of fact, we can show that plasticity is a phenomenon driven by shear stresses and therefore corresponds to Tresca or Von Mises criteria.

At first approximation, we can explain this behavior difference between the brittle materials and the ductile materials by the atomic bond type which makes them up:

– a brittle material is generally made up of covalent bonds, meaning strong bonds which are very stable and directional, coming from bringing together the electrons from amongst neighbor atoms. In order to attain the rupture, these bonds therefore need to be broken. It is therefore necessary to provide a significant normal stress. This reasoning applies well in the instance of pure crystals, but not so well in that of poly-crystals (meaning when the material is made up of different crystals with different orientation which are separated by boundary grains);

– a ductile material is generally comprised of metallic bonds, meaning delocalized strong bonds coming from the bringing together of electrons from amongst a large number of atoms (which then become ions). These electrons make up a rather mobile mist in the material (hence their particularly strong electrical conductivity). The material may therefore be strained by displacing its atoms step by step (the atoms settle for changing neighbors which is not possible, or is at least very difficult, for a covalent bond). This is the dislocation phenomenon. It is therefore necessary to provide a significant shear to enable the displacements of atoms and therefore, the plastic (and irreversible) strain of the material.

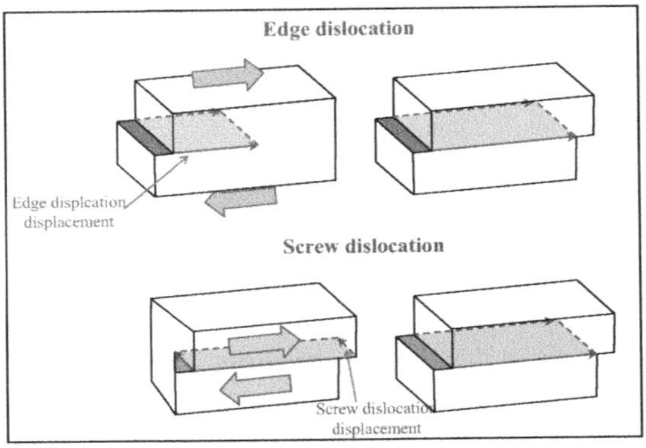

Figure 6.6. *Edge dislocation displacement and screw dislocation under the action of a shear stress. For a color version of this figure, see www.iste.co.uk/bouvet/aeronautical.zip*

6.4. Stress or maximum shear energy criterion: ductile material

6.4.1. *Tresca criterion*

In order to account for the significance of the shear stress on the plasticity of ductile materials, the simplest criterion is the maximum shear criterion or the Tresca criterion. It states that the plasticity of the material can be attained if the maximum shear stress (for all facets or all possible normal vector $\underline{n}$) attains a limit stress.

As it is simpler to perform a tension test, in practice, we prefer to define an equivalent stress, called the Tresca stress, which is equal to the tension stress under a tension applied force. In tension, we have:

$$\underline{\underline{\sigma}} = \begin{bmatrix} \sigma & 0 & 0 \\ 0 & 0 & 0 \\ 0 & 0 & 0 \end{bmatrix}_{(\underline{x},\underline{y},\underline{z})} \qquad [6.9]$$

And we therefore obtain a maximum shear stress (either using Mohr's circle or by finding the maximum of τ for all the possible normal vector $\underline{n}$):

$$\tau_{\max} = \frac{\sigma}{2} \qquad [6.10]$$

We therefore define the Tresca equivalent stress by:

$$\sigma_{tresca} = 2.\tau_{\max} \qquad [6.11]$$

Thus, in pure tension, the Tresca equivalent stress will be directly equal to the tension stress. The Tresca criterion is therefore written as:

$$\sigma_{tresca} = 2.\tau_{\max} < \sigma_e \qquad [6.12]$$

where σ_e is the elastic limit of the material. The Tresca criterion is in fact an end of the elasticity criterion (or the beginning of plasticity) and is not a rupture criterion. This criterion therefore makes sense for determining a material's end of elasticity, but makes no sense for a brittle material. Note that the elastic behavior is just the beginning of a ductile material's behavior:

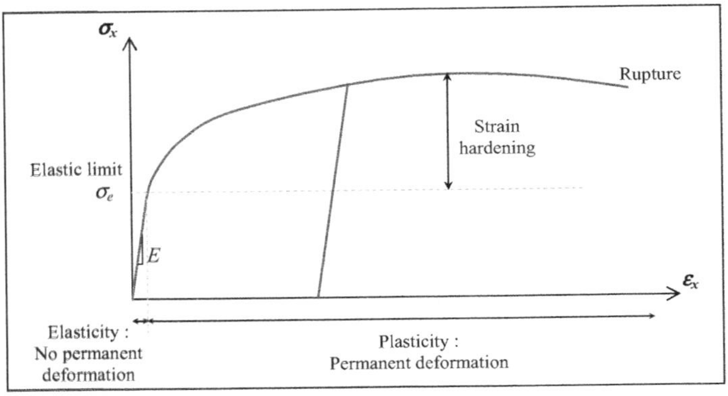

Figure 6.7. *Tension test of a ductile material. For a color version of this figure, see www.iste.co.uk/bouvet/aeronautical.zip*

In 3D, it can also be shown (I will leave you to do the demonstration using Mohr's tri-circle, for example) that the Tresca stress is equal to the maximum difference of principal stresses (in absolute value):

$$\sigma_{Tresca} = Max\left(\left|\sigma_I - \sigma_{II}\right|;\left|\sigma_{II} - \sigma_{III}\right|;\left|\sigma_{III} - \sigma_I\right|\right) \qquad [6.13]$$

In practice, this criterion is not used very often and the Von Mises criterion is preferred, as it is rather similar and it is derivable (see the next paragraph). In fact, if we trace these two criteria in 2D in the principal stress plane (σ_I, σ_{II}), by assuming that they are equal in tension (which is just a hypothesis!), we then obtain:

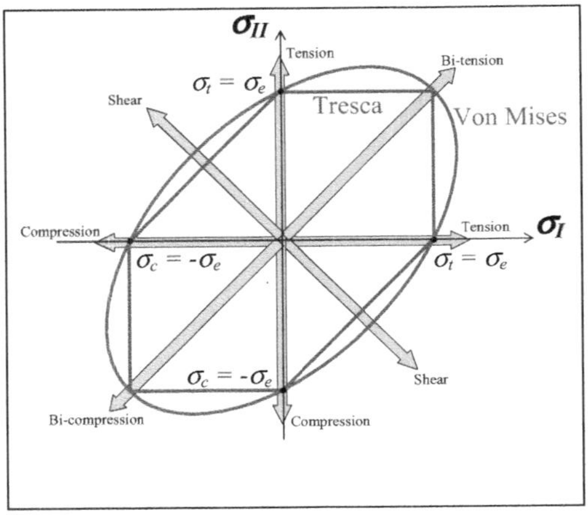

Figure 6.8. *End of elasticity criteria of a ductile material. For a color version of this figure, see www.iste.co.uk/bouvet/aeronautical.zip*

In the Tresca criterion, we observe some angular points which later pose numerical problems for modeling the plasticity.

6.4.2. *Von Mises criterion*

The Von Mises criterion is based on the shear energy. In the same way as the Tresca criterion, this is based on the fact that the plasticity is sensitive to shearing.

The energy corresponds to the energy that the material may store in an elastic form when it is subjected to a given stress and strain. This phenomenon is the same as for a spring which stores energy:

$$E_{ressort} = \frac{1}{2} F.x = \frac{1}{2} k.x^2 \quad [6.14]$$

where F is the force in the spring, k is its rigidity and x is the displacement (with $F = -k.x$).

By analogy, the energy (called the strain energy E_d) of a volume V of matter subjected to a stress $\underline{\underline{\sigma}}$ and to a strain $\underline{\underline{\varepsilon}}$ is half of the stress product by the strain (which is a contracted product here):

$$E_d = \frac{1}{2} \cdot \iiint_V \underline{\underline{\sigma}} : \underline{\underline{\varepsilon}}.dV \quad [6.15]$$

Or, in its more developed form:

$$E_d = \frac{1}{2} \iiint_V \left(\sigma_{xx}.\varepsilon_x + \sigma_{yy}.\varepsilon_y + \sigma_{zz}.\varepsilon_z + 2.\tau_{xy}.\varepsilon_{xy} + 2.\tau_{yz}.\varepsilon_{yz} + 2.\tau_{xz}.\varepsilon_{xz} \right).dV \quad [6.16]$$

We remember that as a matter of fact, a contracted product is the trace of the product or even the equivalent of the scalar product for the vectors:

$$\underline{\underline{\sigma}} : \underline{\underline{\varepsilon}} = trace\left(\underline{\underline{\sigma}}.\underline{\underline{\varepsilon}}\right) = \sum_{i,j} \sigma_{ij}.\varepsilon_{ij} \quad [6.17]$$

Now we need to separate the shear energy from the rest; here the rest corresponds to the dilatation energy or the hydrostatic pressure energy. Hydrostatic pressure corresponds to the stress obtained when a cube is pressed with a (hydro-type) pressure on all the faces of the REV (incidentally, you will be able to show that this tensor is invariant by rotation):

$$\underline{\underline{\sigma}} = \begin{bmatrix} -p & 0 & 0 \\ 0 & -p & 0 \\ 0 & 0 & -p \end{bmatrix} \quad [6.18]$$

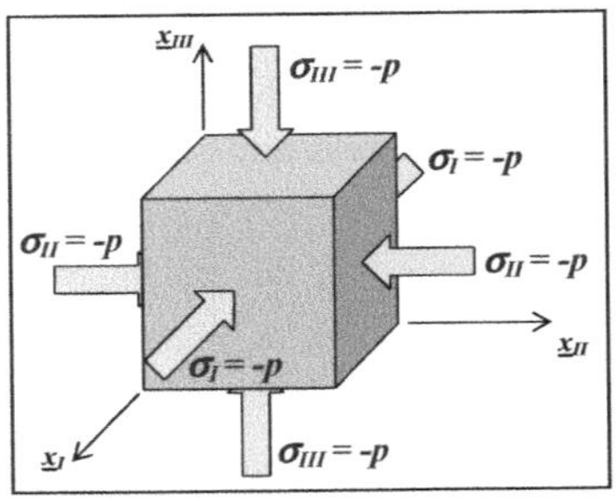

Figure 6.9. *Hydrostatic pressure. For a color version of this figure, see www.iste.co.uk/bouvet/aeronautical.zip*

In fact, we note that plasticity is never reached when a ductile material is subjected to a hydrostatic pressure, even a very significant one. Thus, we can separate the stress tensor into two parts; the hydrostatic part which has no effect on the plasticity and the deviatoric part:

$$\underline{\underline{\sigma}} = \frac{1}{3} trace\left(\underline{\underline{\sigma}}\right).\underline{\underline{I}} + dev\left(\underline{\underline{\sigma}}\right) \quad [6.19]$$

Or if we write the hydrostatic part explicitly:

$$\frac{1}{3} trace\left(\underline{\underline{\sigma}}\right).\underline{\underline{I}} = \frac{1}{3} \begin{bmatrix} \sigma_x + \sigma_y + \sigma_z & 0 & 0 \\ 0 & \sigma_x + \sigma_y + \sigma_z & 0 \\ 0 & 0 & \sigma_x + \sigma_y + \sigma_z \end{bmatrix} \quad [6.20]$$

If we do the same with the strain, the contracted product of the stress tensor and strain tensor is then decomposed into two parts:

– the product of the deviatoric parts of stress and strain;

– the product of the hydrostatic parts of stress and strain.

As the cross products (deviatoric×hydrostatic) are null.

It is this deviatoric part that corresponds to the previously evoked shear energy.

In practice, as the stress and strain are linear amongst themselves, we prefer to use a deviatoric product of the stress tensor and we thus define the Von Mises stress σ_{VM} as the root of this deviatoric product of the stress tensor:

$$\sigma_{VM} = \sqrt{\frac{3}{2}.\underline{dev}(\underline{\underline{\sigma}}):\underline{dev}(\underline{\underline{\sigma}})} \qquad [6.21]$$

As for the 3/2 coefficient, it is just there so that the Von Mises equivalent stress is equal to the tension stress in tension (I will leave you to verify that in pure tension, $\sigma_{VM} = \sigma$).

It can also be written in its developed form:

$$\begin{aligned}\sigma_{VM} &= \sqrt{\frac{1}{2}\cdot\left[(\sigma_x-\sigma_y)^2+(\sigma_y-\sigma_z)^2+(\sigma_z-\sigma_x)^2+6.\left(\tau_{xy}^2+\tau_{yz}^2+\tau_{xz}^2\right)\right]} \\ &= \sqrt{\frac{1}{2}\cdot\left((\sigma_I-\sigma_{II})^2+(\sigma_{II}-\sigma_{III})^2+(\sigma_{III}-\sigma_I)^2\right)}\end{aligned} \qquad [6.22]$$

We therefore consider that the material remains linear elastic if the Von Mises equivalent stress remains lower than the elastic limit σ_e:

$$\sigma_{VM} < \sigma_e \qquad [6.23]$$

In reality, this criterion functions rather well, but you obviously must not forget that in reality, it is the material that decides and the engineer's role is to use the criterion best suited to the studied material!

If we trace the Von Mises criterion in the principal stress plane (σ_I, σ_{II}), we obtain an ellipse passing through $\pm\sigma_e$ in tension/compression (see Figure 6.8).

If we now trace the Von Mises criterion in 3D in the principal stress space $(\sigma_I, \sigma_{II}, \sigma_{III})$, we obtain a circular-based cylinder with a tri-sector axis (the line of equation $x = y = z$) and a radius $\sigma_e/\sqrt{3}$ (the tension axis is not perpendicular to the tri-sector axis, hence the coefficient $\sqrt{3}$). This cylinder translates the fact that the criterion is independent of the hydrostatic pressure, namely of any stress tensor translation according to the tri-sector.

In the same way, we can trace the Tresca criterion which also gives a cylinder, but with a hexagonal base:

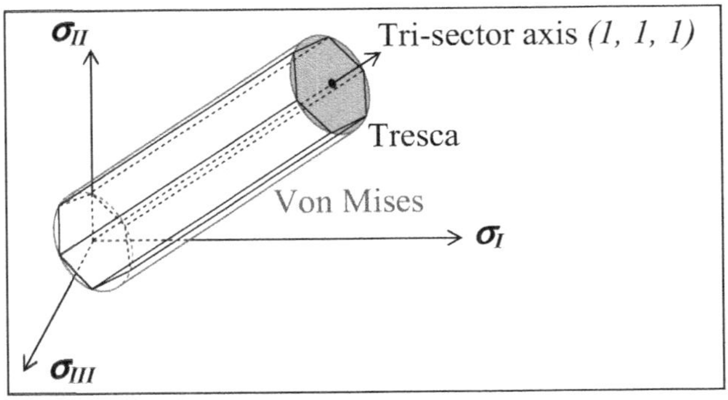

Figure 6.10. *End of elasticity criteria of a ductile material. For a color version of this figure, see www.iste.co.uk/bouvet/aeronautical.zip*

If we now compare the Tresca and Von Mises criteria, in both 2D and 3D, we realize that the difference between these two criteria is rather weak. To choose, we need to experimentally measure the end of elasticity criterion of the studied material and compare these experimental points to the theoretical criteria. In reality, we note that the experimental dispersions are often greater than the difference between these two criteria, and it is difficult to say if one is better than the other. In practice, we use the Von Mises criterion instead, so as to avoid the angular points of the Tresca criterion.

6.4.3. *Rupture of a ductile material*

If we now carry out a tension test on a ductile material, such as steel, for example, until a rupture appears, we note that the rupture facies is (in general) oriented at 45°. This result is rendered properly by the Tresca and Von Mises criteria which assume the predominance of the shear stress over the normal stress. In fact, we can show that in pure tension, the maximum shear is obtained at 45° of the tension direction (and is worth half of the tension stress).

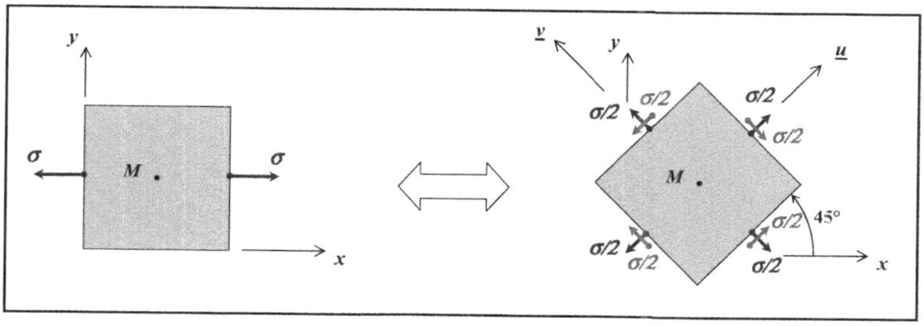

Figure 6.11. *Maximum shear in a tension test. For a color version of this figure, see www.iste.co.uk/bouvet/aeronautical.zip*

Nevertheless, this result needs to be taken with caution, as these criteria allow us to determine the beginning of plasticity, yet the rupture takes place at the end of plasticity. Moreover, it enables us to become aware of the beginning of plasticity, rather than the rupture phenomenon. We can still, however, use a very hard steel, meaning one which has been subjected to heat (and/or mechanical treatment), which enables its elastic resistance to be increased and therefore, in return, its plastic area to be diminished. It is therefore shown that the rupture is definitely at 45°, as predicted by the Tresca and Von Mises criteria. We will also incidentally note that as the shear sign is of no significance (unlike that of the normal stress), the rupture will, indifferently, take place at +45° or at –45° in 2D (and in any plane at 45° of the load axis in 3D).

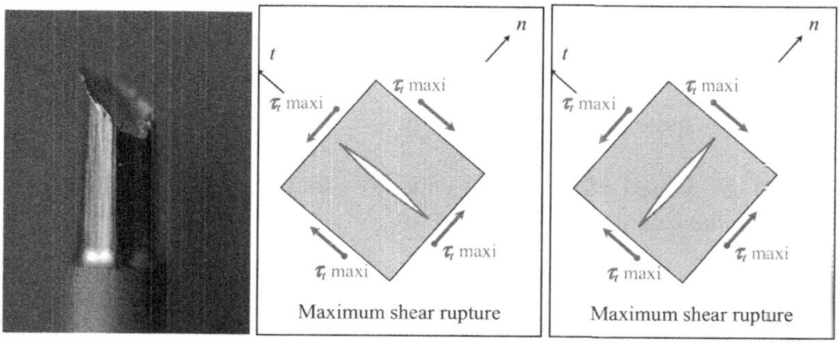

Figure 6.12. *Rupture of a ductile material with little plasticity (Source: https://fr.wikiversity.org/wiki/Introduction_à_la_science_des_matériaux). For a color version of this figure, see www.iste.co.uk/bouvet/aeronautical.zip*

This result is less true on a ductile material with significant plasticity, particularly as a significant necking develops throughout the test. The behavior is therefore not at all elastic and the modeling of this behavior is much more complex.

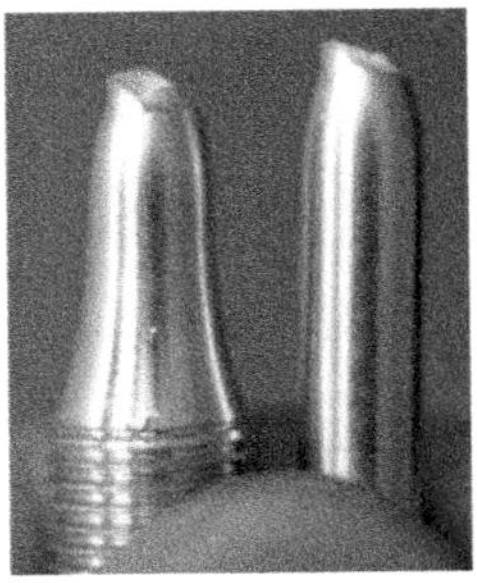

Figure 6.13. *Rupture of a ductile material with much plasticity (Source: https://fr.wikiversity.org/wiki/Introduction_à_la_science_des_matériaux). For a color version of this figure, see www.iste.co.uk/bouvet/aeronautical.zip*

EXAMPLE: SHEARING.–

If we use the Von Mises criterion, we can show that for a pure shear state (I will leave you to do the calculations):

$$\overline{\sigma} = \sqrt{3}.\tau < \sigma_e \quad [6.24]$$

From this, we can also deduce that using the Von Mises criterion automatically leads to:

$$\tau_e = \frac{\sigma_e}{\sqrt{3}} \quad [6.25]$$

which is generally quite well verified.

If we use the Tresca criterion, we can show that for a pure shear state (I will leave you to do the calculation):

$$\sigma_{tresca} = 2.\tau < \sigma_e \quad [6.26]$$

From this, we can also deduce that using the Tresca criterion automatically leads to using:

$$\tau_e = \frac{\sigma_e}{2} \quad [6.27]$$

It therefore remains to choose between the Von Mises coefficient √3 and the Tresca coefficient 2! Even if we prefer to use the Von Mises criterion more often for the reasons conveyed previously, in practice, it is obviously the material that decides, because the difference between these two coefficients is often drowned out by other experimental errors. Indeed, it must not be forgotton that experimentally determining the elastic limits of a ductile material is more complex than simply determining its elasticity modulus.

EXAMPLE: HYDROSTATIC PRESSURE.–

If we consider a material with an applied force by a hydrostatic pressure:

$$\underline{\underline{\sigma}} = \begin{bmatrix} -p & 0 & 0 \\ 0 & -p & 0 \\ 0 & 0 & -p \end{bmatrix} \qquad [6.28]$$

Then we can show that:

$$\sigma_{VM} = \sigma_{tresca} = 0 \qquad [6.29]$$

In short, regardless of the value of p, and even if it tends to infinity, plasticity can never be provoked! Be careful: that does not mean that there is no strain; elasticity is not based on equivalent stresses, and we can still calculate them:

$$\underline{\underline{\varepsilon}} = \begin{bmatrix} \frac{-(1-2.\nu).p}{E} & 0 & 0 \\ 0 & \frac{-(1-2.\nu).p}{E} & 0 \\ 0 & 0 & \frac{-(1-2.\nu).p}{E} \end{bmatrix}_{(\underline{x},\underline{y},\underline{z})} \qquad [6.30]$$

As paradoxical as it may seem, it is properly verified! To do the test, it "suffices" to put a small cube in a liquid tank under pressure and to increase the pressure, note that the strain observed is effectively elastic (or almost).

6.5. Maximum shear criterion with friction: compression of brittle materials

We have seen previously that the maximum principal stress criterion enables us to simulate the rupture of brittle materials at first approximation. This is quite true in tension, but much less so in compression.

In particular, if we subject a brittle material to pure compression, we observe a crack with an angle between 45° and 60° of the loading direction. For example, let us take the case of a carbon fiber-based composite material. This material is evidently anisotropic, as the rigidity (and the rupture stress) in the fiber direction (longitudinal direction *l*) is much greater than that of the resin direction (transversal direction *t* and out-of-plane direction *z*). In contrast, in the plane (*t*, *z*), this material is isotropic. We are therefore talking about transverse isotropy as it presents a symmetry axis (*l*). We can therefore study the rupture of this material in 2D in the plane (*t*, *z*) as that of an isotropic material.

Before going back to the compression, note that in pure tension, this material presents a brittle rupture which starts at the fiber/resin interfaces. The crack is perpendicular to the loading direction, underlining the fact that this material is particularly sensitive to the normal tension stress, which is consistent with the criterion in maximum principal stress.

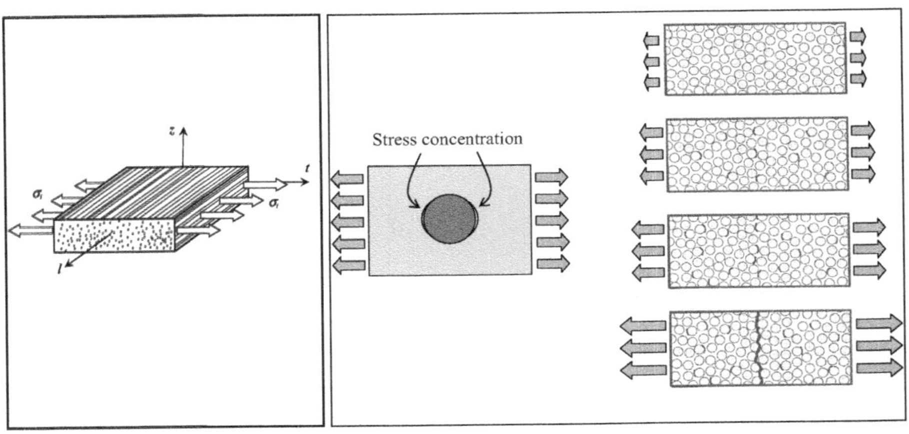

Figure 6.14. *Rupture of a composite under transverse tension. For a color version of this figure, see www.iste.co.uk/bouvet/aeronautical.zip*

Compression is more complex. As a matter of fact, as conveyed previously, we observe a crack making an angle between about 45° and 60° of the loading direction.

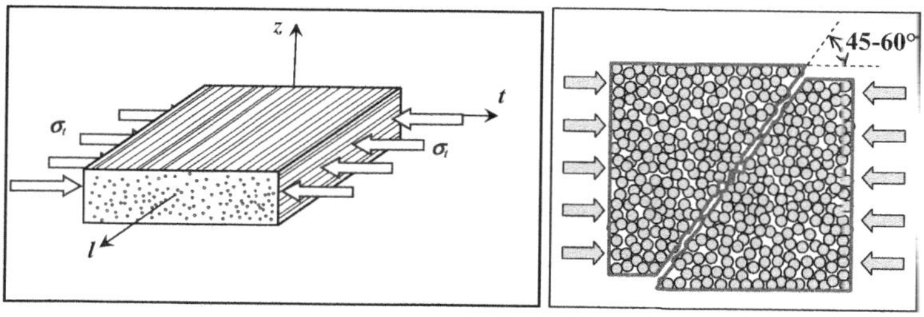

Figure 6.15. *Rupture of a composite in transverse compression. For a color version of this figure, see www.iste.co.uk/bouvet/aeronautical.zip*

At first approximation, we could think that, in the absence of a positive principal stress, the rupture would be due to a shear rupture. If this had been the case, then we would have observed a crack at 45° (the maximum shearing direction is at 45° of the compression direction).

Mohr–Coulomb presented the hypothesis that this rupture is definitely a shear rupture, but that it was upset by the friction due to the negative normal stress.

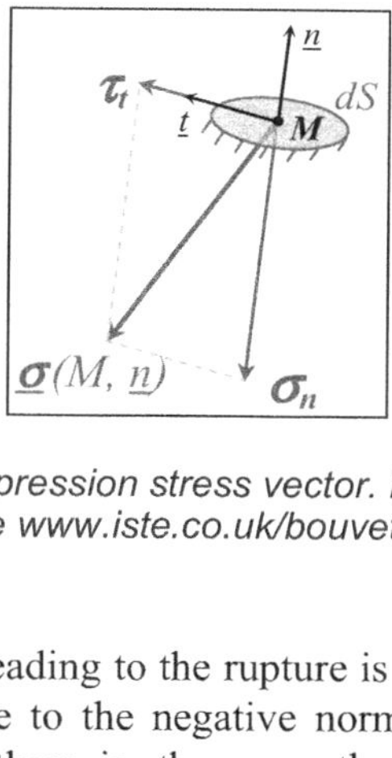

Figure 6.16. *Compression stress vector. For a color version of this figure, see www.iste.co.uk/bouvet/aeronautical.zip*

In other words, the stress leading to the rupture is definitely a shear stress, but is diminished by the friction due to the negative normal stress. In other words, the more negative normal stress there is, the more the material can withstand shear stress before breaking. Thus, we obtain:

$$|\tau_t| + f.\sigma_n \leq \tau_0 \qquad [6.31]$$

where $f = tan\ (\phi)$ is the friction coefficient, σ_n is the normal stress (here $\sigma_n \leq 0$) and τ_t is the shear stress.

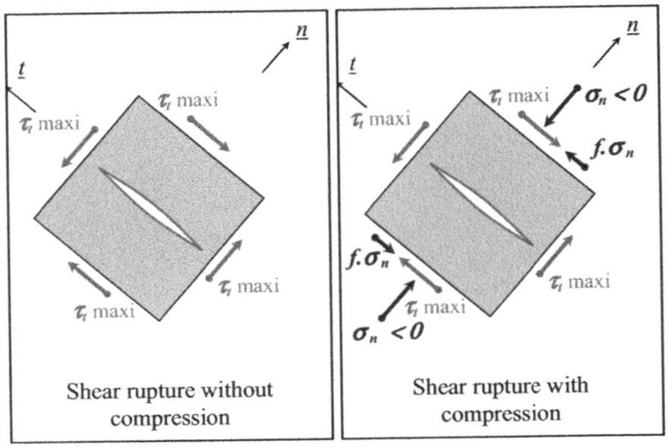

Figure 6.17. *Shear rupture without and with compression. For a color version of this figure, see www.iste.co.uk/bouvet/aeronautical.zip*

In the absence of a normal stress (or if the friction coefficient is null), the criterion is equivalent to the Tresca criterion and we therefore find that the shear limit τ_0 is half of the tension limit:

$$\tau_0 = \frac{\sigma_0}{2} \quad [6.32]$$

where we note σ_0, and not σ_e, as this criterion is a rupture criterion for a brittle material and does not make any sense for ductile materials and therefore, for determining the end of elasticity (except by taking a null friction coefficient, which goes back to using the Tresca criterion!).

We can therefore show that under pure compression, the stress leading to rupture will be greater (in absolute value) than that of tension:

$$\sigma_c = \frac{-\sigma_t.\cos(\varphi)}{1-\sin(\varphi)} \quad [6.33]$$

with σ_t being the tension limit and σ_c being the compression limit (considered to be negative).

In the case where $\sigma_I \geq \sigma_{II} \geq \sigma_{III}$, we can then show that the Mohr–Coulomb criterion can be written in the following form:

$$\frac{1+.\sin(\varphi)}{1-\sin(\varphi)}\sigma_I - \sigma_{III} \leq -\sigma_c \quad [6.34]$$

This criterion is especially used in civil engineering for studying concretes, rocks or soils.

A specific case is that of pulverulent materials, meaning powder materials or materials which crumble and transform into powder very quickly, as is the case for certain sands and soils. A pulverulent material therefore presents a null (or very weak) tension limit (and thus, a null shear limit). The only way to make it withstand the shear (which is the case for ground surrounding a building's foundations) is therefore to subject it to compression stress, which the previous equation renders well.

If we trace the Mohr–Coulomb criterion in the principal stress plane (σ_I, σ_{II}), just like for the Tresca criterion, we almost obtain a hexagon with a dissymmetry in tension/compression.

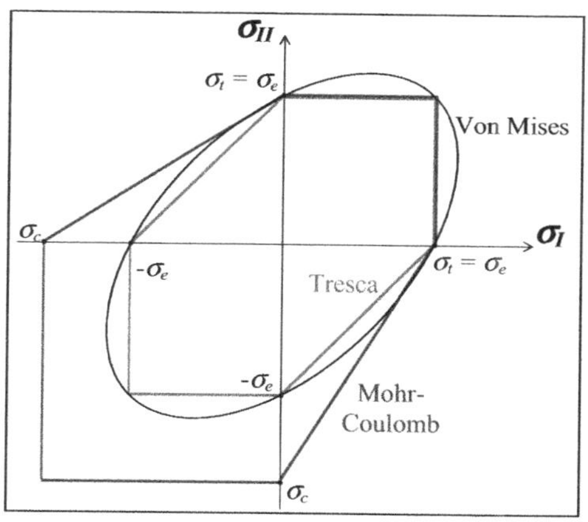

Figure 6.18. *Mohr–Coulomb, Von Mises and Tresca criteria. For a color version of this figure, see www.iste.co.uk/bouvet/aeronautical.zip*

If we now trace the Mohr–Coulomb criterion in 3D in the principal stress space (σ_I, σ_{II}, σ_{III}), we obtain a strained cone with a hexagonal base (caused by the tension/compression dissymmetry). This cone form translates the influence of hydrostatic pressure on the Mohr–Coulomb criterion (unlike the Tresca criterion). Physically, this means that a positive hydrostatic pressure leads to rupture, contrary to a negative hydrostatic stress which settles for compressing the material without ever breaking it. This behavior is sufficiently verified for brittle materials. It also corresponds rather well to common sense.

If we now go back to the previously evoked carbon/epoxy compression test, we can then show that this criterion foresees a rupture angle θ_r:

$$\theta_r = \frac{\pi}{4} + \frac{\varphi}{2} \qquad [6.35]$$

If the friction coefficient is null, we definitely find a crack at 45°, and this angle increases with friction. In the case of carbon/epoxy, the rupture angle is about 54°, which would imply a friction angle of 18° and a friction coefficient of 0.32, which is in the order of magnitude of the measured values (or even a little greater).

In contrast, if we look at the ratio between the compression stress and the tension stress, we note that the theory foresees a maximum (in absolute value) of 2.4 for a friction coefficient of 1 (which is already very significant). In reality, this ratio attains about 4 for the carbon/epoxy (50 MPa in tension for –200 MPa in compression)!

Nevertheless, you must bear in mind that this friction concept is very limited before the rupture. As a matter of fact, if there is no rupture, there is no crack and there are no two surfaces in contact like there would be for classic friction. But in our carbon/epoxy case, we can think instead that this coefficient 4 is due to very different rupture phenomena for tension and compression. In tension, the rupture is in fact due to the coalescence of cracks created on the fiber/resin interfaces; cracks created as a cause of significant stress concentration (the fiber and resin materials have different rigidity). However, in compression, as the rupture is due to the shear, there is no reason for the stress concentration coefficient to be the same in tension, and must likely be lower.

As always, you must bear in mind that a model remains just that: a model. The reality is always more complex than the small model into which we wish it to fit! In other words, always remember to look properly at the reality before blindly using a model.

EXAMPLE: CALCULATION OF THE MOHR-COULOMB CRITERION IN COMPRESSION.–

If we consider a pure compression state in 2D (we can show that the result is the same in 3D):

$$\underline{\underline{\sigma}} = \begin{bmatrix} \sigma & 0 \\ 0 & 0 \end{bmatrix} \qquad [6.36]$$

with, of course, $\sigma < 0$. For a facet with normal vector $\underline{n}$ oriented at an angle θ of the x direction, we obtain:

$$\begin{cases} \sigma_n = \sigma.\sin^2(\theta) \\ \tau_t = \sigma.\sin(\theta).\cos(\theta) \end{cases} \quad [6.37]$$

Hence:

$$\sigma.\left(f.\sin^2(\theta) - \sin(\theta).\cos(\theta)\right) \leq \frac{\sigma_0}{2} \quad [6.38]$$

All that remains it to determine for which value θ_r of θ this stress is maximum, meaning after calculation:

$$\tan(2\theta_r) = \frac{-1}{f} \quad [6.39]$$

Hence:

$$\theta_r = \frac{\pi}{4} + \frac{\varphi}{2} \quad [6.40]$$

And the compression stress is subsequently:

$$\sigma_c = \frac{-\sigma_t}{2.\left(\sin(\theta_r).\cos(\theta_r) - f.\sin^2(\theta_r)\right)} = \frac{-\sigma_t.\cos(\varphi)}{1 - \sin(\varphi)} \quad [6.41]$$

6.6. Anisotropic criterion: example of the composite

We now go back to the composite example presented previously. We have seen that this material has an anisotropic elastic behavior:

$$\begin{cases} \varepsilon_l = \dfrac{\sigma_l}{E_l} - \dfrac{\nu_{lt}.\sigma_t}{E_l} \\ \varepsilon_t = \dfrac{\sigma_t}{E_t} - \dfrac{\nu_{lt}.\sigma_l}{E_l} \\ \varepsilon_{lt} = \dfrac{\tau_{lt}}{2.G_{lt}} \end{cases} \quad [6.42]$$

At first approximation, we can show that this material can be considered brittle. We can easily grasp that its rupture criterion is not isotropic, but anisotropic. Moreover, after having seen the complex rupture phenomena at play, we can show that a tension/compression dissymmetry is present.

In order to simplify the problem, we are going to place it in 2D in the plane (l, t).

To account for the different behaviors observed, we can use a principal stress criterion, taking care to separate the different directions:

$$\begin{cases} \sigma_l^c \leq \sigma_l \leq \sigma_l^t \\ \sigma_t^c \leq \sigma_t \leq \sigma_t^t \\ |\tau_{lt}| \leq \tau_{lt}^r \end{cases} \quad \text{with} \quad \begin{cases} \sigma_l^c \leq 0 \\ \sigma_t^c \leq 0 \end{cases} \qquad [6.43]$$

with σ_l^t and σ_l^c the tension and compression stresses at rupture in the longitudinal direction, σ_t^t and σ_t^c the tension and compression stresses at rupture in the transverse direction and τ_{lt}^r the stress at shear rupture. For example, for the previously evoked T300/914, we have:

$$\begin{cases} \sigma_l^t = 1500 \ MPa \\ \sigma_l^c = -1400 \ MPa \\ \sigma_t^t = 50 \ MPa \\ \sigma_t^c = -200 \ MPa \\ \tau_{lt}^r = 75 \ MPa \end{cases} \qquad [6.44]$$

As mentioned previously, the drawback of this model is that it does not account for coupling between the directions. Therefore, we most commonly use the Tsai–Hill criterion.

This criterion is based on a generalization of the Von Mises criterion. The Von Mises criterion is written in 2D:

$$\sigma_{VM} = \sqrt{\sigma_l^2 + \sigma_t^2 - \sigma_l.\sigma_t + 3.\tau_{lt}^2} \leq \sigma_e \qquad [6.45]$$

Tsai–Hill have then generalized it in the following form:

$$A.\sigma_l^2 + B.\sigma_t^2 + C.\sigma_l.\sigma_t + D.\tau_{lt}^2 \leq 1 \qquad [6.46]$$

And we can easily show that in the case of tension in the longitudinal direction, we have:

$$A = \frac{1}{(\sigma_l^{\,t})^2} \qquad [6.47]$$

In the case of tension in the transversal direction, we have:

$$B = \frac{1}{(\sigma_t^{\,t})^2} \qquad [6.48]$$

And in pure shear:

$$D = \frac{1}{(\tau_{lt}^{\,r})^2} \qquad [6.49]$$

The last parameter C is determined by considering that in bi-axial tension ($\sigma_l = \sigma_t$ and $\tau_{lt} = 0$), the rupture will be attained when σ_l is attained σ_l^t, hence we can show that:

$$C = \frac{-1}{(\sigma_l^{\,t})^2} \qquad [6.50]$$

In practice, this hypothesis is difficult to verify as bi-axial tests on this type of orthotropic material with a large difference in behavior between the directions l and t is rather delicate to interpret. Nevertheless, as the influence of this parameter C remains rather weak in practice, in comparison to other parameters, its value barely modifies the criterion. The Tsai–Hill criterion is therefore written as:

$$\left(\frac{\sigma_l}{\sigma_l^{\pm}}\right)^2 + \left(\frac{\sigma_t}{\sigma_t^{\pm}}\right)^2 - \frac{\sigma_l.\sigma_t}{\left(\sigma_l^{\pm}\right)^2} + \left(\frac{\tau_{lt}}{\tau_{lt}^{\,r}}\right)^2 \leq 1 \qquad [6.51]$$

With:

$$\begin{cases} \sigma_l^{\pm} = \sigma_l^{\,t} & if \quad \sigma_l > 0 \\ \sigma_l^{\pm} = \sigma_l^{\,c} & if \quad \sigma_l < 0 \\ \sigma_t^{\pm} = \sigma_t^{\,t} & if \quad \sigma_t > 0 \\ \sigma_t^{\pm} = \sigma_t^{\,c} & if \quad \sigma_t < 0 \end{cases} \qquad [6.52]$$

This can therefore be seen as a sum of elastic energies (divided by the rupture stress) in each of the directions, which enables us to partially account for the interaction between the different damages.

If we trace this criterion in 2D in the plane (σ_l, σ_t), we obtain a curve made up of four ellipse pieces. In this figure, the maximum stress criterion has been shown in order to draw a comparison.

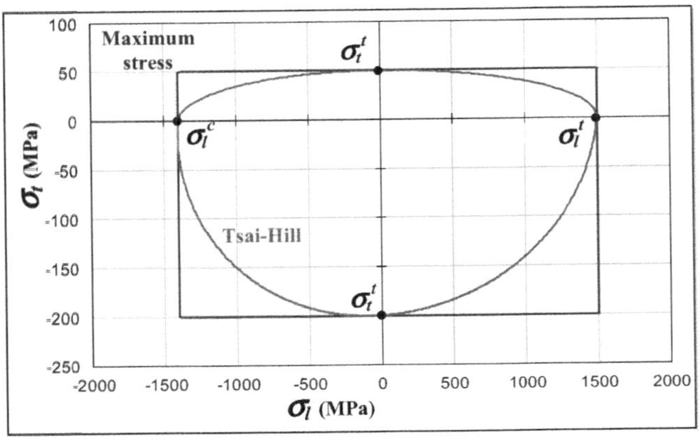

Figure 6.19. *Tsai–Hill and maximum stress criteria. For a color version of this figure, see www.iste.co.uk/bouvet/aeronautical.zip*

7

Plasticity

7.1. Introduction

Now that you have seen the main criteria for sizing a structure, we are going to look at what happens after these criteria.

For brittle materials, we generally observe a brutal rupture of the structure just after the criterion. The criteria that we have just studied (maximum normal stress, Tsai–Hill or Mohr–Coulomb depending on the materials studied) can therefore be used as both an end of elasticity criterion and a rupture criterion. Nevertheless, there are a certain number of materials which have more complex behavior, such as composite materials, for example, for which the rupture is more or less progressive. In this instance, the Tsai–Hill criterion generally enables us to determine the first ply rupture, but that does not generally provoke a rupture of all the other plies at once.

For ductile materials, we generally observe plasticity developing once the criterion has been attained. In fact, if we observe the tension curve of a ductile material, we observe an elastic linear beginning followed by a plastic part (meaning that a permanent strain exists when the stress is released). Therefore, the criteria that we have just studied (Tresca or Von Mises) cannot be used as an end of elasticity criterion and a rupture criterion at the same time. These criteria enable us to determine the end of elasticity, but other approaches will be necessary to determine the rupture. In fact, using these criteria to determine the rupture would again overlook the plasticity part in the tension curve (yet the plasticity represents the largest part of this curve)!

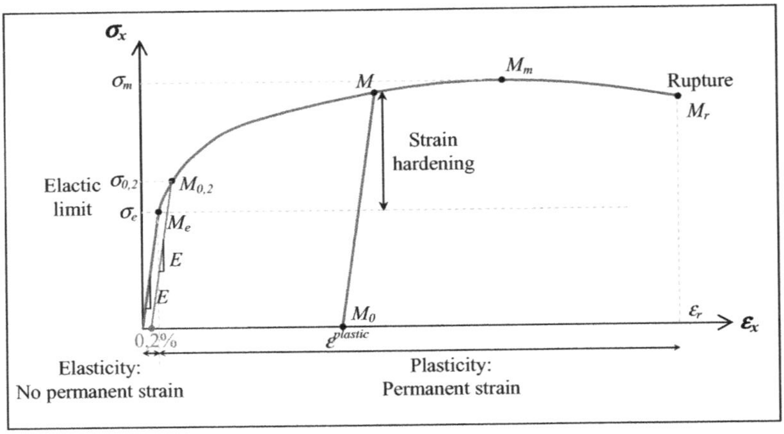

Figure 7.1. *Plasticity of a ductile material. For a color version of this figure, see www.iste.co.uk/bouvet/aeronautical.zip*

This plastic behavior (and the plastic strain in particular) is very significant for a structure's behavior. For example, due to this plastic strain, we can shape or stamp pieces (and that they do not return to their initial sheet shape).

Furthermore, due to this plastic strain, we can increase the elastic limit of a material by strain hardening. In fact, if we load our material up to point M and if we unload it up to the null stress, we then obtain a material with an equilibrium point M_0 and with an elastic limit σ_M (greater than the initial elastic limit σ_e): we are now talking about strain hardening. Incidentally, we will note that the new equilibrium position M_0 will therefore be seen as a null strain for the "new" material and that the stress/strain curve will be defined as based on M_0. You must also bear in mind that in reality, given the manufacturing process, and in particular the heat treatment, a piece must necessarily be in a strain hardening state, and that only the M_0 state is known.

If the strain hardening enables the material's elastic limit to be increased, it also induces its toughness and strain at rupture to be decreased. In fact, we note that the stress/strain curve of the "new" material presents a weaker plasticity domain, a lower strain at rupture ε_r and a weaker dissipated energy (corresponding to the air under the curve). This capacity that the material has to dissipate energy is directly connected to the fracture toughness (or to the critical stress intensity factor): the more the material can dissipate energy by plastic deformation, the more significant the fracture toughness will be, and the more slowly the cracks will propagate (and *vice versa*). We find the same opposition as for the precipitation hardening (see. Chapter 8): increasing the elastic limit decreases the fracture toughness (and *vice versa*).

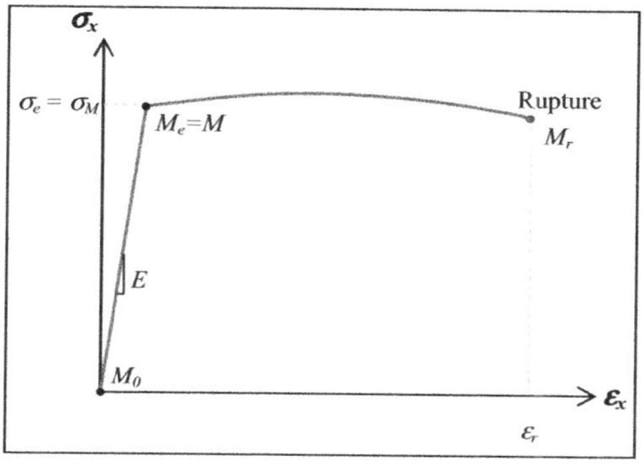

Figure 7.2. *Plasticity of a ductile material after prior strain hardening. For a color version of this figure, see www.iste.co.uk/bouvet/aeronautical.zip*

In practice, it is almost impossible to define an elastic limit σ_e as the passage between elasticity and plasticity is done in a (more or less, depending on the material) very progressive way. Thus, we define an elastic limit at 0.2% ($\sigma_{0.2}$) corresponding to the point inducing a plastic strain of 0.2%. In practice, (due to malapropism) we often call this 0.2% stress "the elastic limit", which we note as σ_e.

Plasticity is also very significant for sizing aeronautical structure, in particular under UL. In fact, we have seen that an airplane must withstand the LL without becoming permanently deformed: the stress must in fact remain elastic, meaning less than the elastic limit. However, for the UL, only one catastrophic rupture is forbidden: a permanent (or plastic) deformation is therefore permitted. An aeronautical structure is therefore sized under the UL, while accounting for the plasticity (or at least the beginning of plasticity).

7.2. Plastic instability: necking, true stress and true strain

If we now observe the shape of the sample in a tension test, we note a plastic strain localization phenomenon which starts at the maximum stress point and which is called necking.

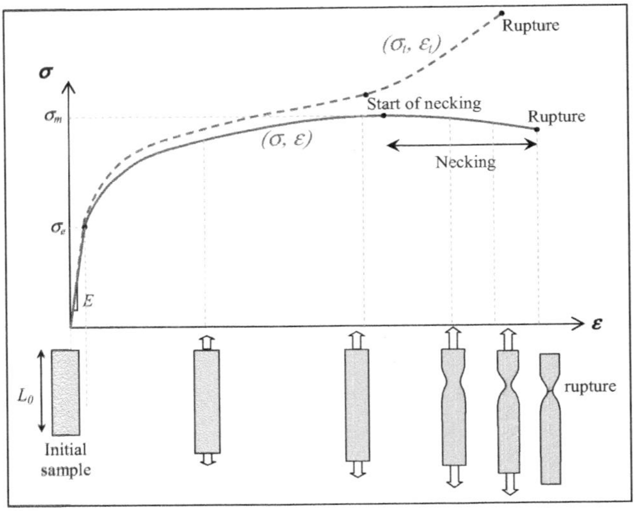

Figure 7.3. *Tension test in engineering and true stress/strain. For a color version of this figure, see www.iste.co.uk/bouvet/aeronautical.zip*

In order to understand this necking, we must study the evolution of the stress depending on the strain, and also the evolution of the sample section. The force of a sample section is obviously the product of this stress by the section:

$$F = \sigma.S_0 \quad [7.1]$$

As long as this force increases with the strain, the sample will deform in a homogeneous way. But if this force sets about decreasing (or remaining constant), then a strain localization, meaning a necking, will appear. In practice, there must be a defect somewhere in the sample where the necking starts, and as the problem is unstable, the necking will become even more localized.

In practice, this necking is a domain in which the structure never works (except in the event of an accident during the final rupture). In fact, this instability is a very dangerous domain for the structural integrity. It is not used for shaping or stamping either, as this would provoke non-homogeneous deformation and the piece would not have the desired shape.

In practice, necking strain on ductile materials occurs in the order of magnitude of a few tens of percentages and about 15% for a 2024 aluminum alloy.

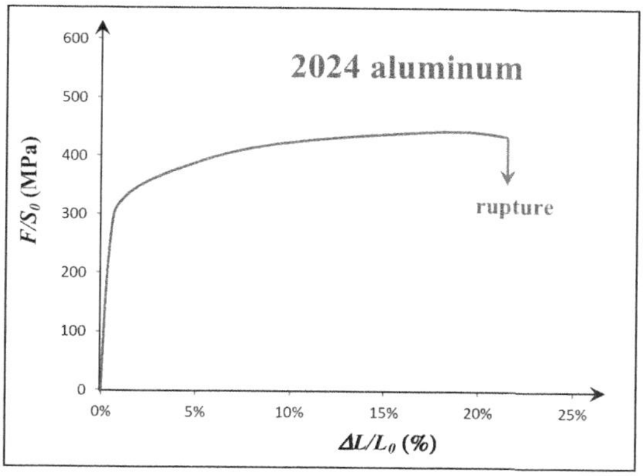

Figure 7.4. *Tension test of a 2024 aluminum alloy. For a color version of this figure, see www.iste.co.uk/bouvet/aeronautical.zip*

From this necking point onwards, as the stress and strain are no longer homogeneous in the sample, the stress/strain tension curve no longer makes sense. In some areas of the sample, there is less stress and strain, and in other areas (such as the necking area), there are more. We can even show that the stress state in the necking area is no longer a pure tension state, but a tri-axial stress state.

This is why, when working on the previous curve, we took care to note F/S_0 and $\Delta L/L_0$ instead of σ and ε. In fact, even after necking, this curve makes sense, but must be seen as the curve of a structure (at least after necking), and not as a material curve.

In reality, even before necking, this curve poses a problem with regards to the definitions of stress and strain adopted. Indeed, these stress and strain are evaluated against the reference position (or L_0 and S_0 here). It is only true for small strain (and small displacement), which in practical terms, is a few percentage points. In the case of plasticity, this strain is quickly exceeded and it is necessary to account for the variation in geometry. In other words, if L_0 is no longer the length of the sample, it makes no sense to divide by L_0!

In referring to the real length L and the real section S, the true stress and true strain (or, more precisely, the strain increment), are defined thus:

$$\begin{cases} d\varepsilon_t = \dfrac{dL}{L} \\ \sigma_t = \dfrac{F}{S} \end{cases} \quad [7.2]$$

As for the previously evoked strain and stress, ε and σ, they are often called engineering stress and strain.

The length L which evolves over time, the strain must in fact be defined incrementally, meaning by integrating:

$$\varepsilon_t = \int_{L_0}^{L} \frac{dL}{L} = \ln\left(\frac{L}{L_0}\right) = \ln\left(1 + \frac{\Delta L}{L_0}\right) = \ln(1+\varepsilon) \qquad [7.3]$$

For the definition of the true stress, we must first of all make a hypothesis about the variation of S, based on a hypothesis about the volume variation.

We have seen that in pure tension, the strain is worth:

$$\underline{\underline{\sigma}} = \begin{bmatrix} \sigma & 0 & 0 \\ 0 & 0 & 0 \\ 0 & 0 & 0 \end{bmatrix} \Rightarrow \underline{\underline{\varepsilon}} = \begin{bmatrix} \varepsilon & 0 & 0 \\ 0 & -\nu.\varepsilon & 0 \\ 0 & 0 & -\nu.\varepsilon \end{bmatrix} \text{ with } \varepsilon = \frac{\sigma}{E} \qquad [7.4]$$

The volume variation is therefore:

$$\frac{\Delta V}{V} = trace\left(\underline{\underline{\varepsilon}}\right) = (1 - 2.\nu).\varepsilon \qquad [7.5]$$

As in general, ν is in the order of magnitude of 0.3, the volume variation in tension is therefore positive. In other words, the volume increase in the direction of tension is not compensated enough by the narrowing of the sample section.

All of that is true in elasticity, but in plasticity, it is not at all true! We in fact observe that the volume variation in plasticity is almost null.

Physically, this behavioral difference between elasticity and plasticity is well understood by the strain phenomena at play. In elasticity, the strain is due to the lengthening of the atomic bonds. The bonds are stretched in the direction of tension applied force, all while shortening these bonds by a small amount in the perpendicular directions and the volume increases. In plasticity, the strain is due to dislocation movements which settle for displacing the crystalline network defects and that do not provoke any variation to the volume. This is incidentally why plasticity is not sensitive to hydrostatic pressure: in fact, hydrostatic pressure tends to decrease (or increase, depending on its sign) the volume of the material, yet, as plasticity has no effect on this volume, the two phenomena do not interact with each other.

In plasticity, the plastic strain will therefore be in the form below:

$$\underline{\underline{\varepsilon}}^P = \begin{bmatrix} \varepsilon^P & 0 & 0 \\ 0 & -\frac{1}{2}.\varepsilon^P & 0 \\ 0 & 0 & -\frac{1}{2}.\varepsilon^P \end{bmatrix} \qquad [7.6]$$

And therefore, whatever the value of ε^P may be, the volume variation will be null.

If we now write that a volume V, with an initial value V_0, does not vary throughout the loading, we obtain:

$$V = V_0 \text{ with } \begin{cases} V_0 = S_0.L_0 \\ V = S.L = S.L_0.(1+\varepsilon) \end{cases} \qquad [7.7]$$

where L_0 (S_0) is the initial length (section) of the volume in the tension direction and L (S) is its length (section) at the instant in question.

Evidently, this is only true if the volume does not vary. In other words, this is not true in elasticity! However, because in practice, plastic strain is much greater than elastic strain (once elasticity threshold has been exceeded), the volume is assumed to be constant. Thus, we obtain a section S:

$$S = \frac{S_0}{1+\varepsilon} \qquad [7.8]$$

The section definitely decreases, while ε is positive.

And thus, the true stress and strain are worth:

$$\begin{cases} \varepsilon_t = \ln(1+\varepsilon) \\ \sigma_t = \sigma.(1+\varepsilon) \end{cases} \qquad [7.9]$$

If we now transform the engineering stress/strain curve in true stress/strain, we generally obtain a curve which strictly increases up until the final rupture. In reality, you must bear in mind that the spirit of the curve part after the necking, makes no sense. In fact, this part of the curve makes even less sense in engineering stress/strain as, in this instance, we are able to consider it as a structural curve, whereas in the case of true stress and strain, even true stress and strain do not make sense after necking. Indeed, we can define a true stress/true strain curve up until the rupture, but the problem becomes a true structural problem and the hypotheses that we have just made to define these true stress and strain are no longer all valid.

7.3. Plastic behavior law: Ramberg–Osgood law

In order to model a test which makes plasticity appear, it is necessary to adapt the elastic behavior law seen previously. To do so, we start by assuming that the strain can be seen as a sum of elastic strain ε^e and of plastic strain ε^p:

$$\varepsilon = \varepsilon^e + \varepsilon^p \quad [7.10]$$

The elastic (linear isotropic homogeneous) behavior law is still valid, but only between stress and elastic strain:

$$\sigma = E.\varepsilon^e = E.(\varepsilon - \varepsilon^p) \quad [7.11]$$

This relation translates a fundamental difference with elasticity: in order to determine the stress, it is insufficient to simply know the stress; the plastic strain must also be known. And to know this plastic strain, it is necessary to know the material's history, such as the evolution of the stress from the beginning, for example.

Practically, we define the evolution of ε^p according to the stress by the intermediary of the strain hardening curve, meaning the threshold stress curve, depending on the plastic strain:

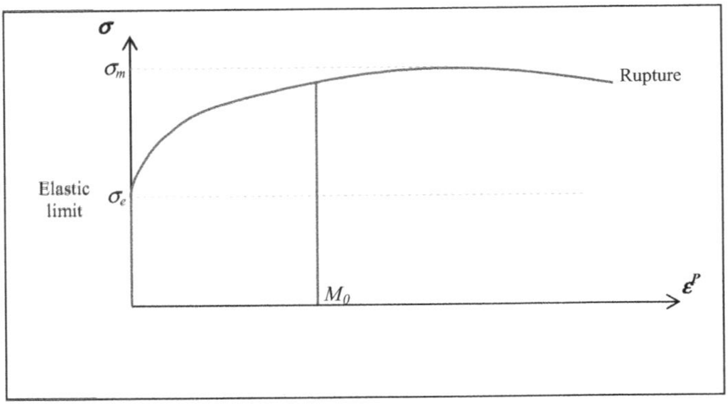

Figure 7.5. *Plasticity curve of stress versus plastic strain. For a color version of this figure, see www.iste.co.uk/bouvet/aeronautical.zip*

In fact, this curve only has one threshold and beyond which, the behavior remains elastic. If the stress remains less than this threshold, for a given plastic strain, the behavior will remain elastic. And if the stress exceeds this threshold, then the plastic strain will evolve.

In aeronautics, we often use the Ramberg–Osgood law:

$$\varepsilon = \varepsilon^e + \varepsilon^p = \frac{\sigma}{E} + 0{,}002.\left(\frac{\sigma}{\sigma_{0,2}}\right)^n \quad [7.12]$$

This means that for a stress equal to $\sigma_{0.2}$, the plastic strain is equal to 0.2% (or 0.002).

Here are some characteristic values for the main metallic materials used in aeronautics.

	2024 Aluminum alloy	7075 Aluminum alloy	TA6V Titanium
E (MPa)	70,000	74,000	82,000
$\sigma_{0.2}$ (MPa)	270	420	210
N	7	14	5,4
$\varepsilon_{rupture}$	20%	10%	25%

Table 7.1. *Mechanical characteristics of the main aeronautic metal alloys*

This gives the tension curves plotted in Figure 7.6.

Of course, you must bear in mind that these curves only make sense for strain which is lower than the rupture strain ε_r. In practice, this rupture strain is in fact the necking strain. Indeed, as conveyed previously, the tension curve makes no sense after necking and it is impossible to use the material beyond this strain because of strain localization problem.

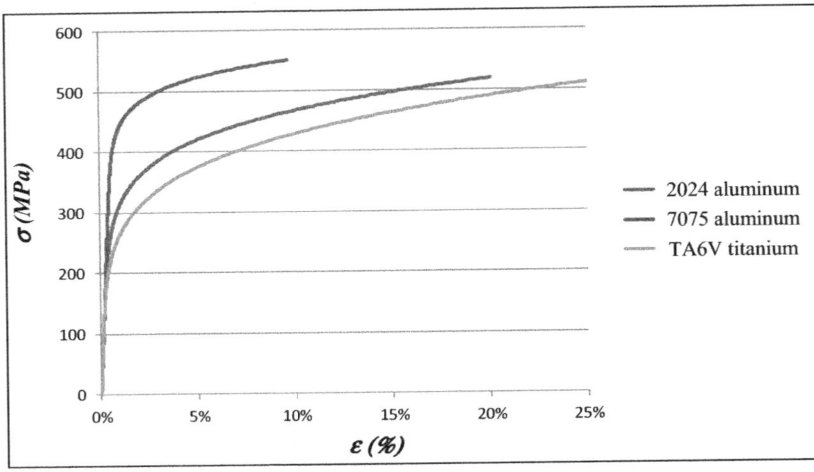

Figure 7.6. *Tension curves of the main aeronautical metal alloys. For a color version of this figure, see www.iste.co.uk/bouvet/aeronautical.zip*

7.4. Example of an elastic–plastic calculation: plate with open hole in tension

In order to illustrate the use of accounting for the plasticity in the structure calculation, we are going to study the example of a holed plate in tension.

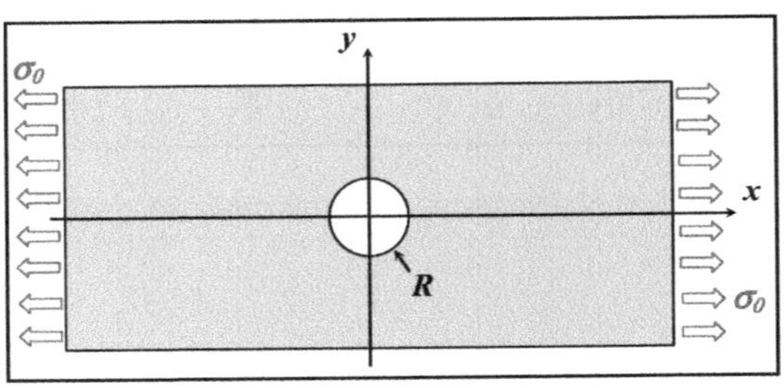

Figure 7.7. *Open hole tension test. For a color version of this figure, see www.iste.co.uk/bouvet/aeronautical.zip*

Incidentally, you can refer to the relevant exercise at the end of this course and show that it is possible to analytically determine the stress in this plate if the instance behavior is linear elastic isotropic homogeneous and if the plate dimensions are very large in comparison to those of the hole.

In order to study the problem of a finite plate example and then to compare it with the elastic–plastic example, we have modeled this example using the finite element method:

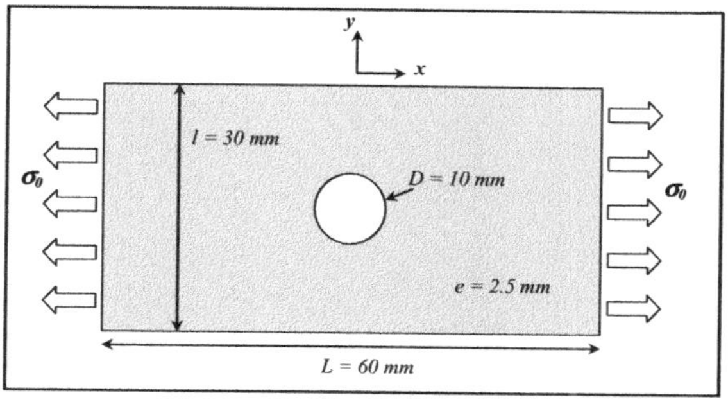

Figure 7.8. *Open hole tension test*

The 10-mm diameter of the hole is akin to a hole with a screw passing through it, and the 30-mm size to the distance between two consecutive screws.

The grid is comprised of membrane-type quadrangular elements and the stress is considered to be plane stress. Only a quarter of the structure is modeled and the boundary conditions are imposed on the axes of symmetry: in $x = 0$, $u = 0$ and in $y = 0$, $v = 0$.

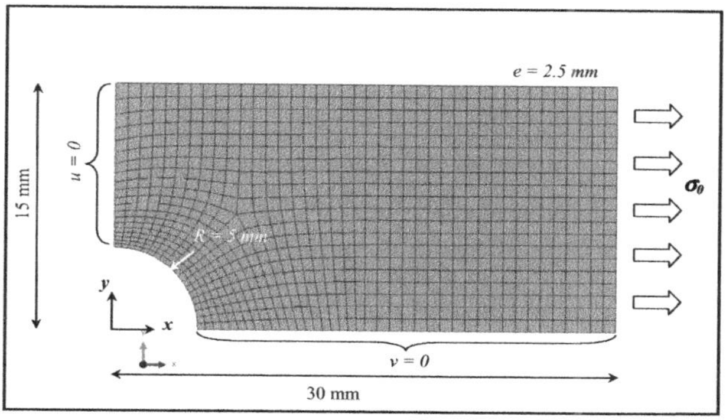

Figure 7.9. *FE modeling of an open hole tension test. For a color version of this figure, see www.iste.co.uk/bouvet/aeronautical.zip*

Considering that the material is aluminum (E = 70 GPa and ν = 0.3), for an average imposed stress of σ_0 = 200 MPa, we then obtain the following Von Mises stress field:

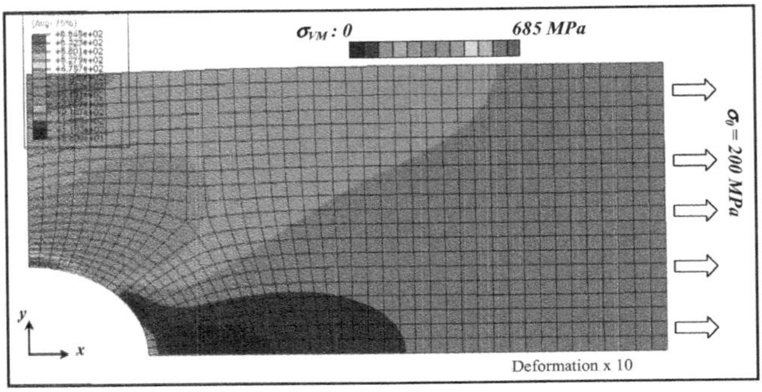

Figure 7.10. *Von Mises stress of an elastic tension test. For a color version of this figure, see www.iste.co.uk/bouvet/aeronautical.zip*

We obviously observe a stress concentration on the edge of the hole with a maximum Von Mises stress of 685 MPa, attained above the hole (x = 0 and y = R), which thus gives a stress concentration factor:

$$K_t^{el} = \frac{\sigma_{VM}^{\max}}{\sigma_0} = 3.4 \qquad [7.13]$$

This coefficient is to be contrasted with the coefficient 3 found in the example of an infinite plate (see example at the end of Chapter 9). In this case, the coefficient is slightly greater as the dimensions of the plate are much greater than those of the hole, which restricts the working section in question (at 2/3 here).

We now consider that the elastic–plastic material, which has an elastic limit stress of 400 MPa, a rupture strain of 20% for a rupture stress of 520 MPa (in fact, these are necking strain and stress) and a linear strain hardening. These characteristics more or less correspond to a standard 2024 aluminum alloy.

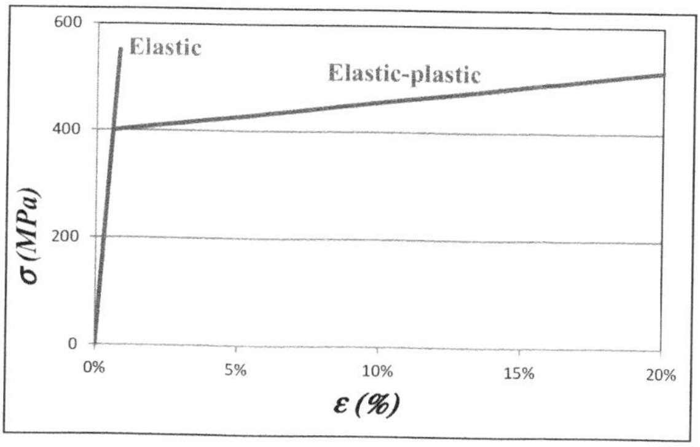

Figure 7.11. *Elastic and elastic–plastic tension curves. For a color version of this figure, see www.iste.co.uk/bouvet/aeronautical.zip*

We therefore obtain the following Von Mises stress field:

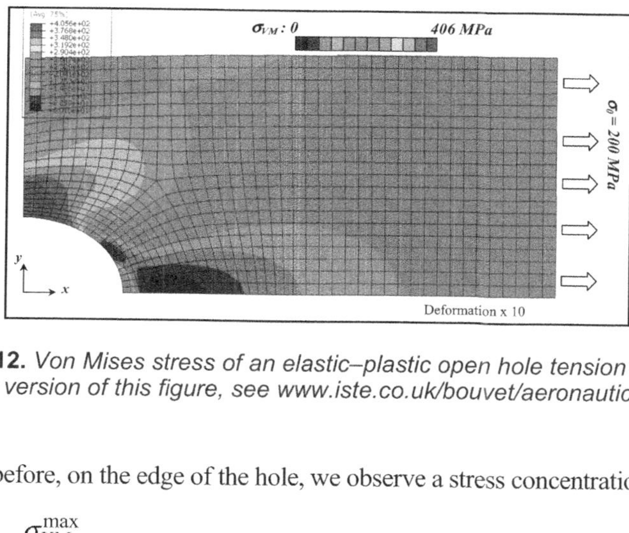

Figure 7.12. *Von Mises stress of an elastic–plastic open hole tension test. For a color version of this figure, see www.iste.co.uk/bouvet/aeronautical.zip*

Just like before, on the edge of the hole, we observe a stress concentration factor of:

$$K_t^{pl} = \frac{\sigma_{VM}^{\max}}{\sigma_0} = 2.0 \qquad [7.14]$$

Thus, the stress concentration factor has decreased from 3.4 to 2.0, simply thanks to the plasticity of the material on the edge of the hole! We now observe the plasticity area. To do so, it suffices to display the plastic strain ε^{pl} (which is obviously worth 0 if the material is in the elastic domain)! We can show that ε^{pl} is a tensor and that just like a stress tensor, we can define a Von Mises equivalent strain (which is equal to the tension strain in pure tension):

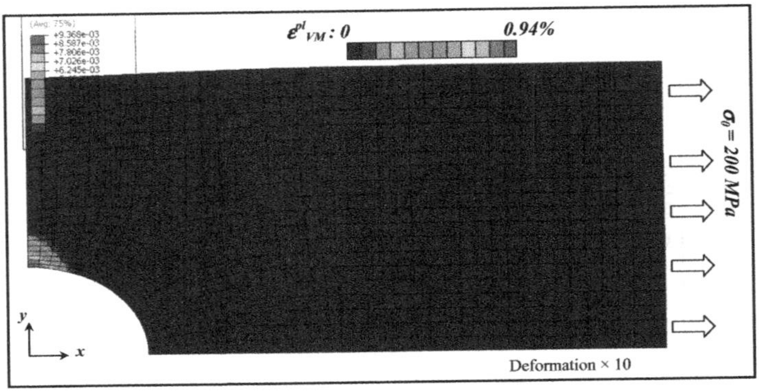

Figure 7.13. *Von Mises plastic strain of an elastic–plastic open hole tension test. For a color version of this figure, see www.iste.co.uk/bouvet/aeronautical.zip*

We note that the plasticity area is very weak and confined to a high stress concentration area. Moreover, the maximum plastic strain remains weak; lower than 1%. This is evident if we compare the material's tension curve and the Von Mises stress field; as the max stress is 406 MPa, the corresponding point is located at the beginning of the plastic area.

We are now going to observe the structure's global mechanical behavior curve. To do so, we trace the average stress σ_0 according to the average strain ε_0, calculated by dividing the displacement imposed at the end of the plate by the length of the plate. Incidentally, we can show that the symmetries used do not change the problem at all – obviously on the condition that the displacement is divided by the half-length, 30 mm, rather than by the entire length, 60 mm!

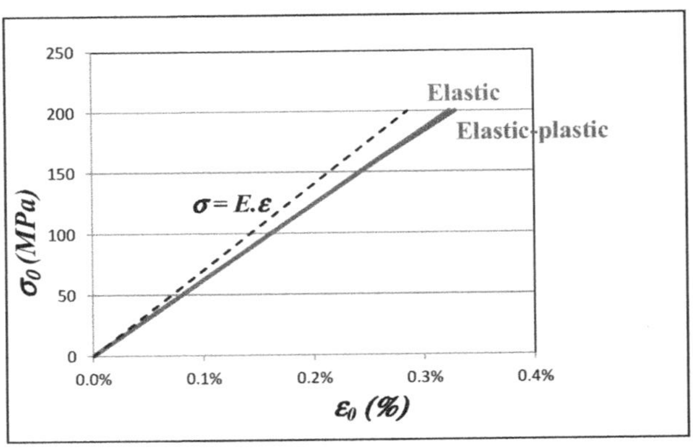

Figure 7.14. *Global curve of an open hole tension test. For a color version of this figure, see www.iste.co.uk/bouvet/aeronautical.zip*

As a comparison, we have also introduced the pure Hooke's law ($\sigma = E.\varepsilon$) into this curve, which would be obtained in the absence of a hole. We note that the loss of rigidity caused by the hole is relatively weak; about 12%.

We then observe that the global behavior of the structure varies very little, irrespective of whether the material is elastic or elastic–plastic; the loss of rigidity is about 1.5%. This is evidently due to the fact that the plastic area remains weak and almost does not affect the global response of the structure.

This example is representative of a defect in the structure. Indeed, when a metal structure, or more simply, when a metallic material piece, is being manufactured, there are necessary locally numerous defects, such as the inclusion of foreign bodies, porosity, metal shrinkage, micro-cracks, etc. When the first loads are applied to the structure, a large stress concentration will then appear around these defects and will therefore generate plastic areas. These plastic areas will have the effect of smoothing the stress and thus, rendering all of these defects much less detrimental than in the absence of plasticity. This is evidently why ductile materials are (generally) less brittle than brittle materials! This capacity of a material to generate plastic areas around the defects can be seen as the fracture toughness of a material, to limit the propagation of a crack; the more plastic areas the material generates, the more energy is needed to create them, and the higher the fracture toughness is. In short, the plasticity must be seen as a positive phenomenon for the holding of the structure, and not as a phenomenon that is to be avoided at all costs!

Certain brittle materials, such as composite materials, for example, develop another phenomenon to smooth the stress around the defects: damage. Damage, which is classically comprised of micro-cracks, indeed requires energy to create this network of micro-cracks and therefore restricts the propagation of cracks and defects; and therefore, increases the material's fracture toughness.

This example of studying the elastic–plastic holed plate must also be contrasted to the LL and UL. In particular, we can ask ourselves the following question: knowing that the plasticity starts to develop a little before 200 MPa, which LL and which UL can this plate withstand? In practice, determining a structure's maximum LL and UL is almost equivalent to a real problem which is, of course, designing a structure, when the LL and the UL are known. Modifying the thickness of the plate, for example, is enough to place the structure's LL and UL to those which are required.

The response to the previous question is less obvious than it may seem. For example, we can place the maximum UL at the level of 200 MPa, saying that we can allow the UL a bit of plasticity, and then place the LL a little below, before the appearance of plasticity. This response evidently meets the LL and UL certification.

In practice, we can even go further by placing the LL at 200 MPa. In fact, although a small plastic area appears, it remains very confined and (almost) does not affect the structure's rigidity, which remains globally elastic! In short, if you do not look very closely, this structure is elastic. Moreover, the loss of elasticity will only appear during the first loading: during the second loading, the material will have local strain hardening and no new plasticity area will appear. Practically, this is how we place the maximum LL of a structure: from the moment when they do not affect the global rigidity to the structure onwards, small plasticity areas are admitted. There is a simple way to allow this and that is to slightly increase the elastic limit of a material being calculated. Thus, this enables us to only perform the elastic calculations, which is simpler and above all, quicker. When looked at closer, this is exactly what is done when the elastic limit used is $\sigma_{0.2}$. Thus, we allow small plastic strains (less than 0.2%) on a local scale, which will allow the stress to be smoothed in the areas of singularity.

To place the UL, we must also study what happens after 200 MPa. To do so, we have gradually increased the load. We then obtain a maximum value of 278 MPa, corresponding to an asymptote of the curve (σ_0, ε_0). From this point onwards, the structure is no longer capable of withstanding the required stress, and the calculation diverges.

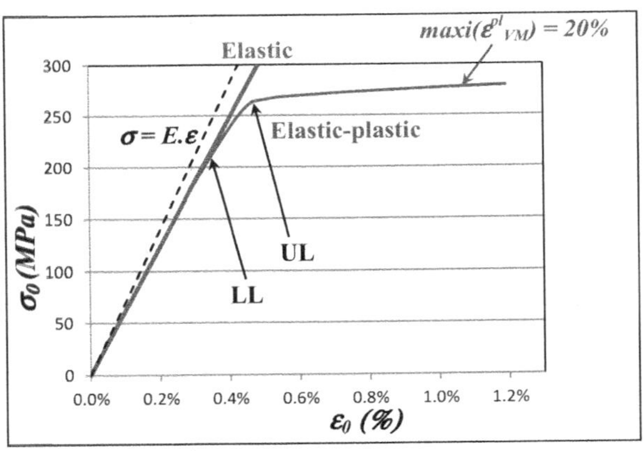

Figure 7.15. *Schematic mechanical behavior of an elastic–plastic structure. For a color version of this figure, see www.iste.co.uk/bouvet/aeronautical.zip*

This curve (σ_0, ε_0) does, nevertheless, allow us to think that it is possible to take an UL of about 260 MPa, which would enable a huge plasticity area to be avoided and to remain below the rupture. As for the LL, it must remain within the order of magnitude of 200 MPa so as to avoid losing any of the structure's rigidity.

In order to understand what happens at 278 MPa, we have traced the Von Mises stress field and the plastic strain field:

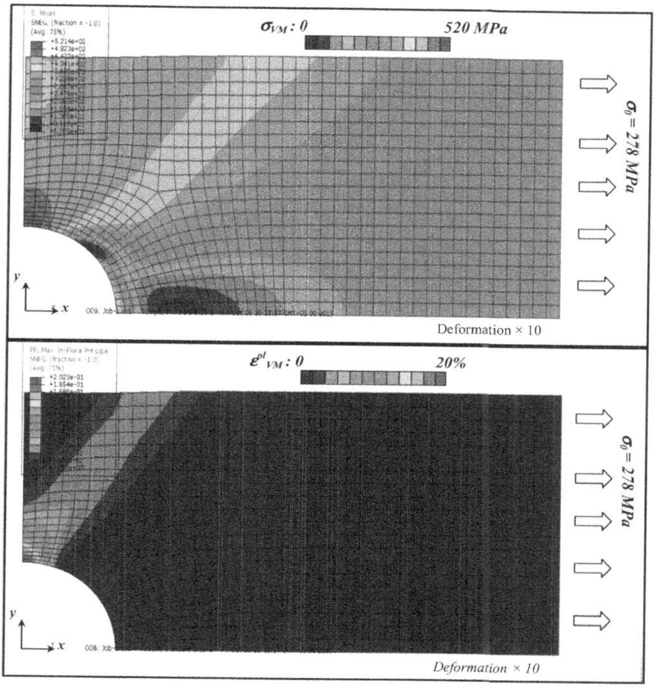

Figure 7.16. *Von Mises plastic stress and strain of an open hole tension test for σ_0 = 278 MPa. For a color version of this figure, see www.iste.co.uk/bouvet/aeronautical.zip*

We evidently observe that the maximum Von Mises stress is equal to 520 MPa and that the maximum plastic strain is equal to 20%, as imposed on the material's curve (σ, ε). If we impose an average stress greater than 278 MPa, all the points which attain 520 MPa of Von Mises stress, break and the structure is therefore no longer capable of withstanding the imposed force.

Nevertheless, this calculation must be taken with caution, as it is false from the moment where a point attains 20% of plastic strain onwards! In fact, from the moment when the most loaded point attains 20% of the plastic strain onwards, the material's behavior cannot be described by the behavior law obtained on the basis of the tension test caused by the appearance of necking. We can verify that in our case, this point is attained just before 278 MPa, and that our model can therefore be considered viable until that point.

8

Physics of Aeronautical Structure Materials

8.1. Introduction

In order to dimension a structure correctly, we must first have a full understanding of the materials in question. This gives us greater perspective as to the behavioral models used (elastic, plastic, viscoelastic, damaged, etc.) together with a better understanding of them. Indeed, we must always be aware that a material's real behavior is invariably more complex than the model. Accordingly, here we will address the two main materials used in aeronautical structures, i.e. aluminum and carbon/epoxy composite. Together, these materials comprise approximately 70% of the structural mass of a commercial aircraft such as the Boeing 787:

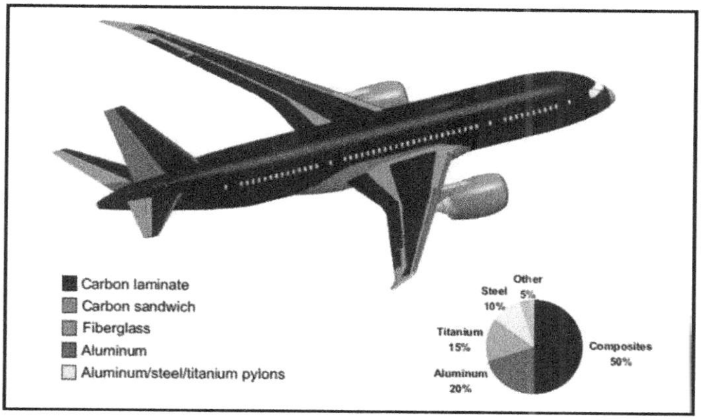

Figure 8.1. *Materials in a Boeing 787 (according to http://www.boeing.fr). For a color version of this figure, see www.iste.co.uk/bouvet/aeronautical.zip*

The other structural materials used are different composite materials based on fiberglass, sandwich structures (a honeycomb core covered by two composite skins), titanium, steel, etc.

However, while we must remember that in aircraft such as the Boeing 787 or the Airbus A350, the proportion of composite structural mass can reach 50%, in more standard aircraft, such as the Airbus A320 or A380, more than 60% of mass is comprised of aluminum alloy.

Steel and titanium are used for highly stressed areas. Titanium offers the advantage of being lighter than steel, but has lower characteristics. We also observe that the most widely used titanium – or to be more precise, the titanium alloy – is an alloy with 6% aluminum and 4% vanadium, called TA6V. Indeed, we will often employ (as a misappropriation) the term aluminum rather than aluminum alloy and titanium rather than titanium alloy.

Material	Composition	ρ (kg/m^3)	E (GPa)	σ_e (MPa)	K_{Ic} (MPa.m^{½})	Use
Ti : TA6V	6%Al, 4%V, etc.	4,450	110	800–1,000	10–13	The most widespread titanium alloy used in aeronautics when aluminum is insufficient
Al : 2024	4.4% Cu, 1.5% Mg, 0.6% Mn, etc.	2,770	70	250–300	30–32	The most widespread aluminum alloy in aeronautics
Al : 2050	3.5% Cu, 1% Li, 0.4% Mg, etc.	2,700	75	350–500	30–35	Aluminum alloy recently used in aeronautics for new programs
Al : 7075	5.5% Zn, 2.5% Mg, 1.5% Cu, etc.	2,800	72	350–500	25–30	Aluminum alloy widely used in aeronautics (more up-to-date than the traditional 2024)
Steel	Fe, 0.02% to 2% C, etc.	7,800	210	200–1,500	30–80	A very broad range between all-purpose steel and maraging steel
Carbon / epoxy	carbon fibers, epoxy resin	1,600	50–150	200–3,000	6–15	The most widespread composite for aeronautical uses

Table 8.1. *Mechanical characteristics of the main materials in aeronautics*

While many aluminum alloys are used in aeronautical structures, here we will limit our scope to three typical aluminum alloys: 2050, 7075 and 2024.

The first number of an aluminum alloy's designation indicates its class. In effect, there are eight classes of aluminum according to their composition:

Series	Composition	Properties	Use
1000	Al not alloyed (content > 99.9%)	Low mechanical properties, good thermal and electrical conductivity, weldable	Common products do not require high mechanical properties
2000	Al-Cu (2024, etc.), Al-Cu-Li (2050, etc.), Al-Cu-Mg, etc.	Good mechanical properties, difficult weldability, poor corrosion resistance	Structures requiring mechanical resistance (particularly in aeronautics)
3000	Al-Mn	Quite good mechanical properties, weldable	As with 1000 series with better mechanical property
4000	Al-Si	Poor mechanical properties, very good castability	Foundry
5000	Al-Mg		Piping, boilers
6000	Al-Mg-Si	Very good weldability, very good formability when hot	Vehicle bodywork, pylons, tubes, welded structures, ship building
7000	Al-Zn-Mg et Al-Zn-Mg-Cu (7075, etc.)	Very good mechanical property	Transport equipment (including aeronautics)
8000	Other alloys of Al		Al-Fe : household packaging

Table 8.2. *Mechanical characteristics of principal aluminum alloys in aeronautics*

2000 and 7000 series aluminum (and particularly 2024, 2050 and 7075) offer excellent static characteristics (E: Young's modulus, σ_e: elastic limit and K_{Ic}: tenacity). Remember that K_{Ic} tenacity represents the critical stress intensity factor, i.e. the stress intensity factor leading to crack propagation. In physical terms this represents the capacity of a material to resist crack propagation; the lower this value, the slower a crack will grow (and vice versa).

At present, the most widely used alloys are 2024 and 7075. Their mechanical characteristics are actually quite similar with a superior elastic limit for 7075 and a better fatigue limit for 2024. Nonetheless, the widespread use of 2024 is due in large part to its earlier development. We should remember that the development of new aeronautical materials is a lengthy (and thereby expensive) process, requiring certification by the authorities prior to in-flight use.

2050 is a more recent alloy that is beginning to supersede traditional alloys. In effect, the addition of lithium allows for a reduction in density and an increase in the Young's modulus compared to traditional alloys of the 2000 or 7000 series that are currently used in aeronautics. This promising alloy allows for weight reduction in aeronautical structures and is now widely used in new Airbus and Boeing programs. It also provides better corrosion resistance than alloy 2024.

We will now study 2024 in detail with particular consideration for the heat treatment required in order to obtain good mechanical characteristics. Indeed, we must remember that mechanical characteristics (particularly σ_e and K_{Ic}) can vary significantly according to the heat treatment. Unfortunately, an increase in σ_e typically induces a decrease of K_{Ic}, and vice versa. In other words, we have to decide between fracture toughness and strength!

8.2. Aluminum 2024

The most precise 2024 composition (% of mass) is as follows:

Al	Cu	Mg	Mn	Fe	Si
Base	3.8–4.9	1.2–1.8	0.3–0.9	≤ 0.5	≤ 0.5

Table 8.3. *Composition of aluminum 2024*

And the heat treatment of a 2024 alloy for aeronautical structures is as follows:

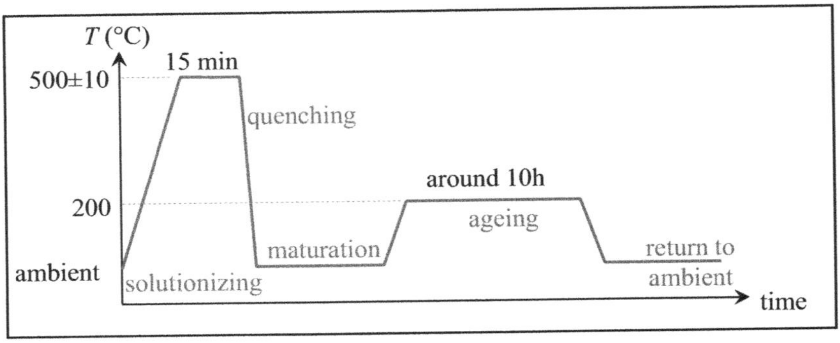

Figure 8.2. *Heat treatment of 2024. For a color version of this figure, see www.iste.co.uk/bouvet/aeronautical.zip*

The first step consists in a "solutionizing treatment". If we are to understand this solutionizing treatment stage, we must first explain the equilibrium diagram for the alloy in question. The equilibrium diagram of a mixture corresponds to the equilibrium states in the mixture according to temperature. Let us consider the Al/Cu diagram for example:

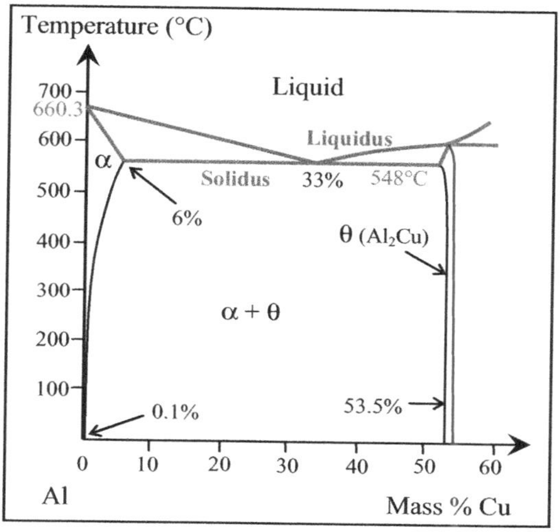

Figure 8.3. *Al/Cu equilibrium diagram. For a color version of this figure, see www.iste.co.uk/bouvet/aeronautical.zip*

In this diagram, the alloy composition figures on the x-axis, which is here the proportion of Cu (with the rest of the 100% being the proportion of Al), and temperature on the y-axis. Accordingly, one point on the diagram allows us to determine the phase of the alloy at a given temperature. In reality, we would also have to consider the effects of other alloying additions, particularly magnesium and manganese, but we can show that the equilibrium diagram does not vary significantly.

If we take this alloy at high temperature (*T > 650°C*, for example at point *A*), the mixture is liquid and composed of a single phase. From 650°C (point *B*), a solid α-phase develops, coexisting with the liquid phase to 580°C (point D). This behavior differs considerably for a pure substance in which the phase-change occurs at constant temperature (with heat exchange occurring relative to the latent heat of the phase-change). For example, this is the case for pure aluminum (0% of Cu in the equilibrium diagram, above) that goes from the liquid to solid phase at 660.3°C, or for water at 0°C (if we account for only regular pressure at atmospheric pressure of course!). The line corresponding to the fusion temperature of an alloy (which varies according to composition) is called liquidus, while the line corresponding to the complete solidification temperature is called solidus.

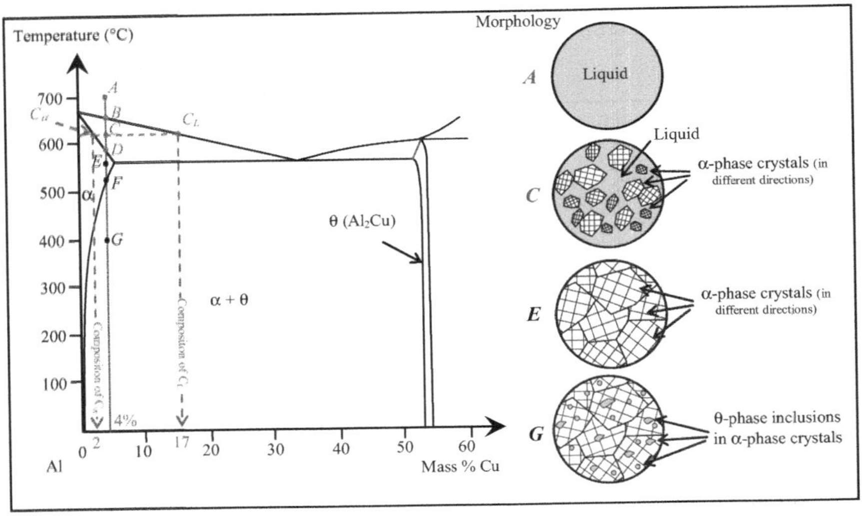

Figure 8.4. *Heat treatment of 2024. For a color version of this figure, see www.iste.co.uk/bouvet/aeronautical.zip*

Between the two phase-change temperatures (points *B* and *D*), we are able to determine the percentage of each phase (α and liquid), along with the composition of each of these phases.

Let us take the example of the 2024 alloy at 620°C (point *C*). We can see in the equilibrium diagram that the liquid phase is composed of 83% Al and 17% Cu (point C_L corresponding to the projection of *C* on the liquidus), and that the solid α-phase is composed of 98% Al and 2% Cu (point C_α corresponding to the projection of *C* on the solidus). The proportion of each of these phases can also be determined by the ratio between segment lengths. The proportion of liquid is equal to $CC_\alpha/C_\alpha C_L$, here 16%, and the proportion of the α-phase is equal to $CC_L/C_\alpha C_L$, here 84%, (this is known as the "inverse segments rule" or the "tie-line rule"). Of course, we can verify that as we are closer to a domain, the more the proportion of the domain's constituent becomes significant; in addition, the total of the two constituents must correspond to 100%. Furthermore, we also see that during alloy cooling, not only do the proportions of the phases present change, but also the composition of each phase changes.

In practice, during cooling the creation of the solid, α-phases occurs throughout the liquid. This process is called nucleation. Typically, nucleation occurs due to defects, for example around the atoms of non-desired elements (an alloy's composition can never be fully perfect, inevitably comprising traces of other elements).

Once point *D* has been passed, the alloy is completely solid, now made up of a single α-phase. In other words, the Al and Cu compounds are entirely miscible, now forming a coherent atomic arrangement. These coherent atomic arrangements occur over small distances, forming the grains of the material. The typical size of a grain is in the order of a tenth of a millimeter. These grains vary from one another in terms of their crystallographic alignment (which, as a general rule, is random). The grains occur from growth at the nucleation sites mentioned above.

Between *D* and *F*, not much happens. If we continue the cooling below point *F*, the α-phase becomes unable to dissolve all the Cu, leading to the formation of a θ-phase (composed of Al_2Cu) in the α-phase. As above, here we can similarly determine the composition of the α- and θ-phases and determine the proportion of each of these phases by means of the inverse segments rule. For example, at 400°C (point G), we obtain 95% of the α-phase and 5% of the θ-phase. In practice, the Cu atoms separate from the α solution in the form of massive inclusions (quite massive precipitates) to form θ-phase inclusions in the α-phase. This phenomenon is made possible by the diffusion of Cu atoms in the α-phase. In practice, this phenomenon is slow and can often be blocked by means of a quenching. A quenching is a rapid cooling obtained by soaking the alloy in water, oil or on occasions simply by means of a cool air flow (the most famous example of quenching would be that of a blacksmith dipping a white-hot sword into water). In the case of quenching, the material obtained is not the mixture as projected by the equilibrium diagram, but a non-equilibrium mixture. Indeed, we should recall that the equilibrium diagram, as its name would suggest, lets us determine the equilibrium state of an alloy at equilibrium point. In practice, alloys are necessarily inclined to develop towards their equilibrium state, but in certain situations this occurs very slowly; in other words, over periods that could indeed be far longer than the structure's use.

Let us now return to the heat treatment of 2024. As in the equilibrium diagram, the solutionizing treatment stage works to create a homogeneous α-phase. The temperature of 500±10°C does not correspond exactly to the Al/Cu equilibrium diagram presented, as the addition of other compounds also modifies this diagram somewhat. The solutionizing treatment duration of about 10 minutes works to dissolve any θ-phase precipitates that could have formed during the material's development (the Cu atoms must be given enough time to diffuse in the α-phase). We should also bear in mind that if the workpiece in question is thick, it will be harder to obtain a homogeneous temperature throughout (thicker or thinner sheet metal).

Next, we perform quenching in water in order to fix the material's structure in this α-phase. This gives us a supersaturated solution, since the proportion of Cu in the α-solution is greater than it should be at ambient temperature. In practice, sheet metal is taken out of the furnace then soaked in a water pool. This procedure must happen quickly in order to avoid cooling and thereby the formation of θ-phase precipitates. We then obtain a material with relatively low hardness and elastic limit (σ_e is in the order of 150 MPa), but with good ductility.

If we keep the alloy at ambient temperature, θ-phase precipitates will form, increasing its hardness and elastic limit (while simultaneously decreasing ductility). In this way, we can obtain an elastic limit up to approximately 300 MPa. These precipitates are much smaller and more numerous than those that could be obtained without quenching. Accordingly, their action is much better, which explains the significant increase in the elastic limit. Note that in order to increase the elastic limit we must restrict the movement of dislocations, by means of precipitates for instance. In practice, while hardening obtained by maturation is of interest, it requires long maturation periods and is often disregarded in favor of a return to an intermediate temperature of around 200°C for several hours.

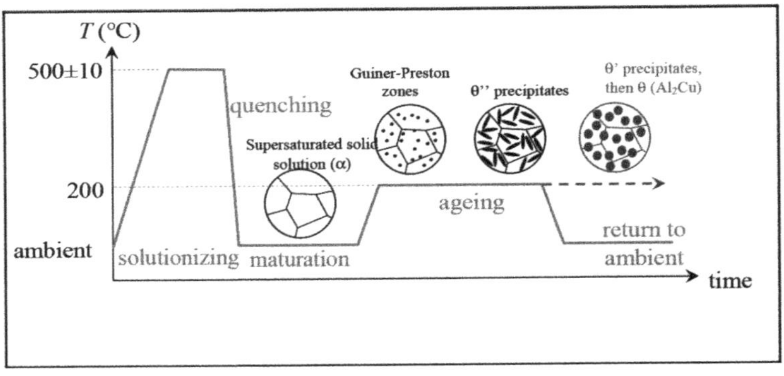

Figure 8.5. *Heat treatment of 2024. For a color version of this figure, see www.iste.co.uk/bouvet/aeronautical.zip*

At the start of aging, the Cu atoms regroup to form aggregates called the Guinier–Preston zone. In θ''-phase, precipitates (this phase has a slightly different composition to that of θ-phase and a different crystallographic structure) are then formed in the shape of platelets. It is by means of these θ'' platelets that we can significantly increase the hardness and elastic limit of 2024 (of course, fracture toughness decreases simultaneously), known as precipitation hardening. If aging is continued, θ'-phase precipitates, allowing for further small gains in hardness. Here, we cease aging for 2024. It is through a judicious mixing of θ' and θ''-phases that we obtain the best elastic limit. If aging were maintained, we would obtain a recomposition of θ' and θ''-phases as θ-phase, and since this phase is more coarse, we would have reduced the elastic limit. Clearly, we see why the 2024 alloy cannot be used at a high temperature: to avoid θ-phase formation. In practice, its use is limited to below 160°C. At higher temperatures, we must use titanium or steel.

The process for hardening aluminum alloys was discovered at the beginning of the 20th Century, though it was only explained by Guinier and Preston in 1938. It was later in 1950, with the development of transmission electron microscopy, that the various θ, θ' and θ''-phases were observed.

This overview of precipitation hardening should give some insight into the complexity accompanying the development of a new alloy. Indeed, metal workers must choose the composition and associated heat treatment in order to obtain the desired mechanical properties at the lowest cost. The possibilities of chemical composition are endless, and the number of heat treatments is similarly infinite; so metal workers' choices are infinite, too!

This section's overview of aluminum 2024 heat treatment is significantly abridged, and the interested reader is referred to [ASH 86], [BRO 04], [DEV 86], [DEQ 12], [DOR 86] and [DUP 13].

8.3. Carbon/epoxy composite T300/914

We will now consider a carbon/epoxy composite material widely used in aeronautical structures: T300/914.

Composite materials are increasingly used in the industry due to their high performance/mass ratio. Of course, this is particularly significant in the aeronautical and space fields due to the crucial importance of the mass criteria for the structures involved. This high performance/mass ratio is due to the use of material with high specific mechanical characteristics, such as carbon fibers, fiberglass or Kevlar.

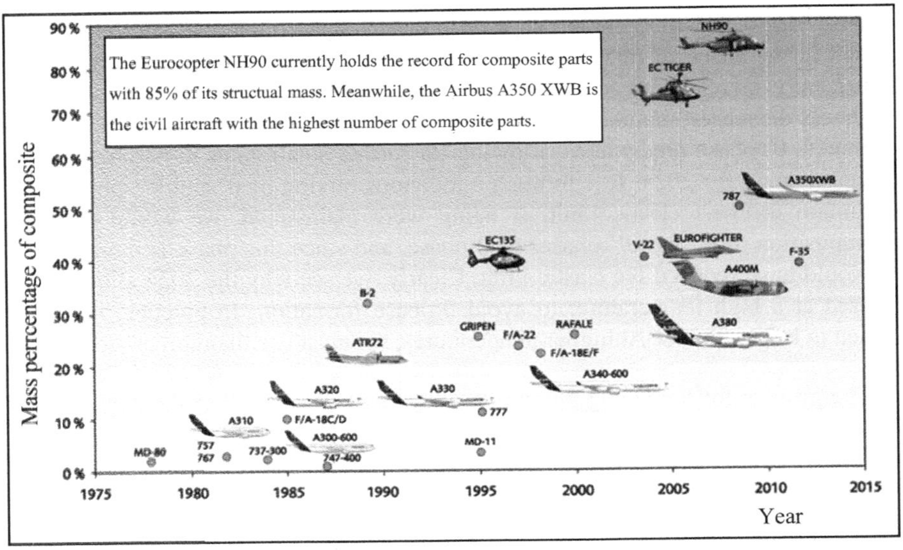

Figure 8.6. *Mass percentage of composite in the structure of Airbus aircraft (according to http://www.airbus.com/). For a color version of this figure, see www.iste.co.uk/bouvet/aeronautical.zip*

Nonetheless, these materials present the major drawback of being more brittle and so must be used alongside less brittle resin-type materials. This is the basic idea of composites, which seek to reinforce a high performance but brittle material, typically in the form of longer or shorter fibers according to the application, by means of a lower performance but less brittle matrix, typically a resin. We should briefly note that though the name would not suggest it, a composite material is not a material but rather the association of at least two materials, typically with complementary characteristics. It should also be noted that on the association of these two materials, an interface between the two is created that also plays its role in overall behavior of the composite.

In particular, we will consider a first generation carbon/epoxy composite made of 50% (in volume) carbon fibers and 50% epoxy resin, traditionally used in aeronautics: T300/914. Where T300 is the reference of the carbon fiber (manufactured by Toray) and 914 that of the epoxy resin. This composite is in the form of a fabric 0.1–0.2 mm thick that can be cut and stacked in order to attain the thickness required.

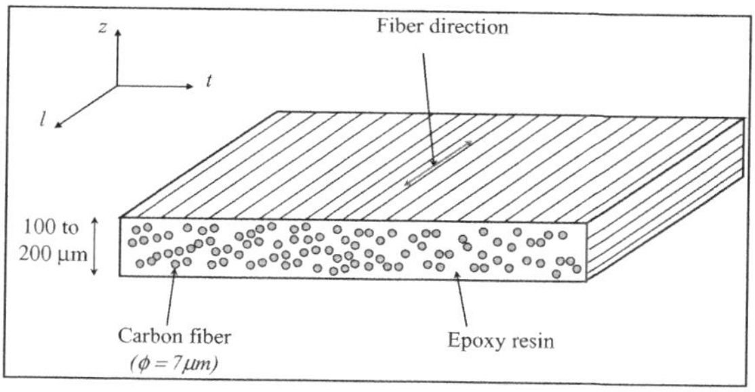

Figure 8.7. *Carbon/epoxy UD lay-up. For a color version of this figure, see www.iste.co.uk/bouvet/aeronautical.zip*

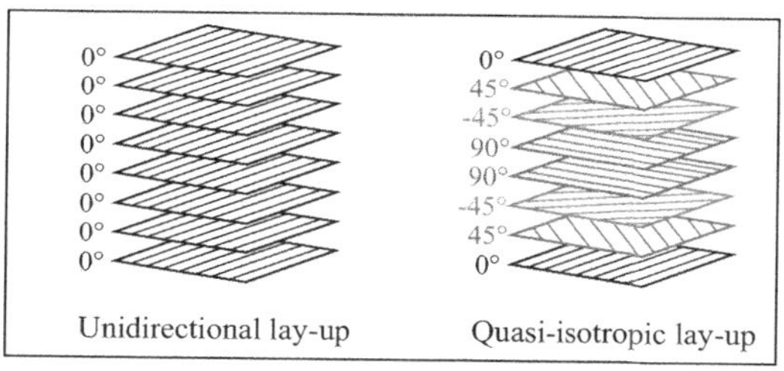

Figure 8.8. *Unidirectional and quasi-isotropic lay-up. For a color version of this figure, see www.iste.co.uk/bouvet/aeronautical.zip*

If we perform a test in the fiber direction, called the longitudinal direction, we observe a brittle elastic behavior similar to that of the fibers. The tensile strength is clearly significantly less than that of the fibers, since around 50% is resin, which has a relatively low tensile strength. This resin is necessary in order to provide a less brittle material and to give the material its form. The carbon fibers are indeed quite interesting features in terms of their mechanical characteristics, but cannot be shaped to allow for the desired geometry. In addition, when cracks occur perpendicularly to the fibers in the material, they provoke numerous fiber cracks and fiber debonding, requiring significant dissipation of energy, making the material less brittle. In practice, such cracks tend to propagate parallel to the fibers if they can, which means that other plies should be put in different directions in order to reinforce the material in all the loading directions (essentially, we can show that the four directions 0°, +45°, −45° and 90° should suffice). Such material is called composite lay-up, as opposed to composite material with a single fiber direction, which is called unidirectional composite.

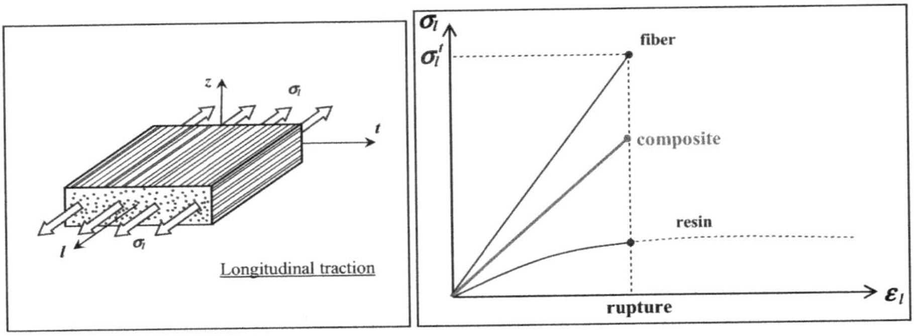

Figure 8.9. *Tension test in longitudinal direction in a composite: fiber, resin and composite behavior. For a color version of this figure, see www.iste.co.uk/bouvet/aeronautical.zip*

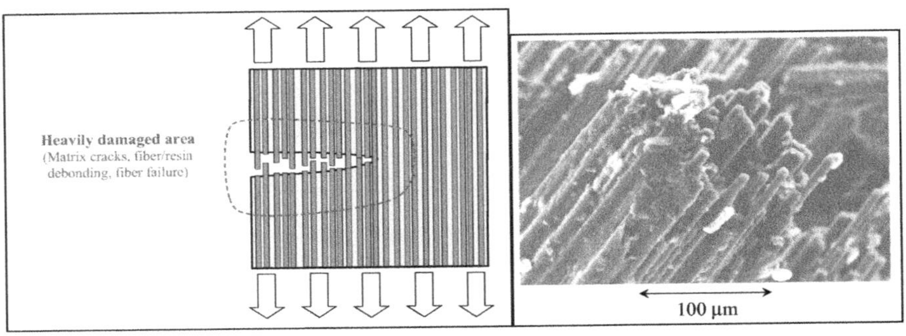

Figure 8.10. *Tension test in longitudinal direction: damaged area. For a color version of this figure, see www.iste.co.uk/bouvet/aeronautical.zip*

The actual structure of a composite is far more complex than that of a more or less homogeneous standard material of the metallic type and requires its own design. The design of a composite structure requires the simultaneous development of a material and a structure; this is the fundamental difference between the development of a metallic structure and a composite structure. In addition to the conventional design iterations of a metallic structure in terms of geometry, composite designs require iterations in terms of the material design; since clearly both of these types of iterations are closely related. In practice, we must add decisions concerning draping sequence or composite manufacturing to the conventional steps of structure geometry.

If we compare Young's modulus and the strength of the main structural materials according to density, we observe that composite materials impress with their performance, particularly when compared to metals. Ceramic materials are also very impressive, despite being too brittle for any structural use.

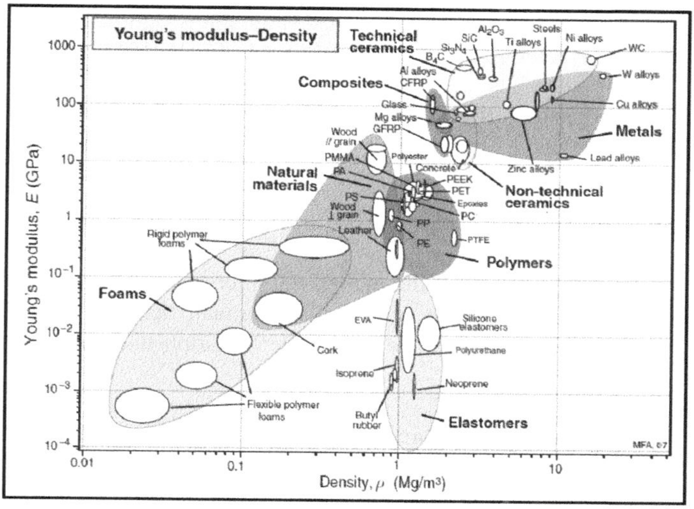

Figure 8.11. *Young's Modulus according to density [ASH 80]. CFRP: Carbon Fiber Reinforced Plastic/GFRP: Glass Fiber Reinforced Plastic. For a color version of this figure, see www.iste.co.uk/bouvet/aeronautical.zip*

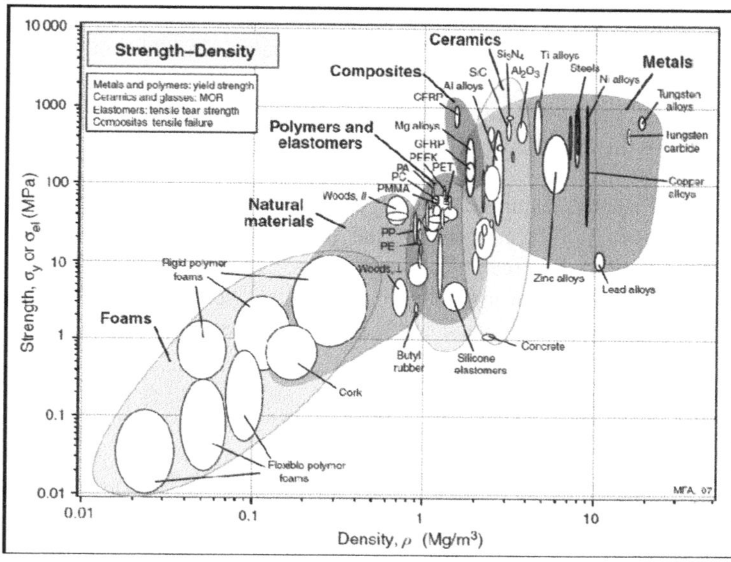

Figure 8.12. *Strength according to density [ASH 80]. CFRP: Carbon Fiber Reinforced Plastic/GFRP: Glass Fiber Reinforced Plastic. For a color version of this figure, see www.iste.co.uk/bouvet/aeronautical.zip*

The overview of T300/914 presented in this paragraph is significantly abridged and the interested reader is referred to [BER 99], [CAS 13], [GOU 10] and [GAY 15]

We will now consider the composition and microstructure of a composite material in detail, and that of epoxy resin in particular.

8.4. Polymers

The epoxy matrix is part of the wider polymer family generally known as plastics. The term "plastic" stems from the mechanical behavior of polymer that often presents plastic deformation, i.e. where deformation does not return to zero when a stress is released.

As the name implies, polymers are composed of monomer chains bonded together by covalent bonds. For present purposes, we will limit our scope to organic polymers. Note that organic matter is made up of various forms of life (vegetables, fungi, animals and microorganisms), and, in particular, through their decomposition. In contrast, inorganic or mineral matter is composed of glass, metal, ceramic or stone.

Organic polymers are based on chains of monomers, themselves based on carbon atoms. The covalent carbon–carbon bond is strong, leading to high mechanical properties. These carbon–carbon bonds act as the basis of the macromolecules that comprise the essential structure of the polymer material. In addition to these strong bonds, these macromolecules are bonded between one another by weak bonds (hydrogen bonds, Van der Waals bonds, etc.). It is the deformation of these weak bonds that induce the significant plastic deformation of polymers.

Let us consider the example of polyethylene, one of the simplest and least expensive polymers. It is composed of the polymerization of ethylene monomers ($CH_2 = CH_2$) leading to the creation of long chains. These chains are bonded between one another simply, by means of strong bonds. Accordingly, the mechanical characteristics obtained remain weak and heavily dependent on temperature.

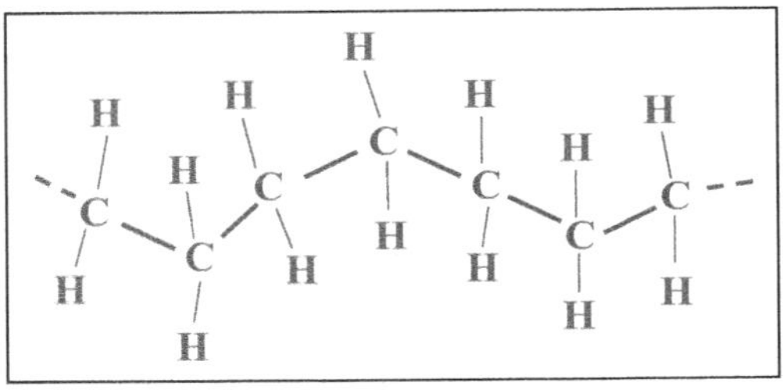

Figure 8.13. *Structure of polyethylene. For a color version of this figure, see www.iste.co.uk/bouvet/aeronautical.zip*

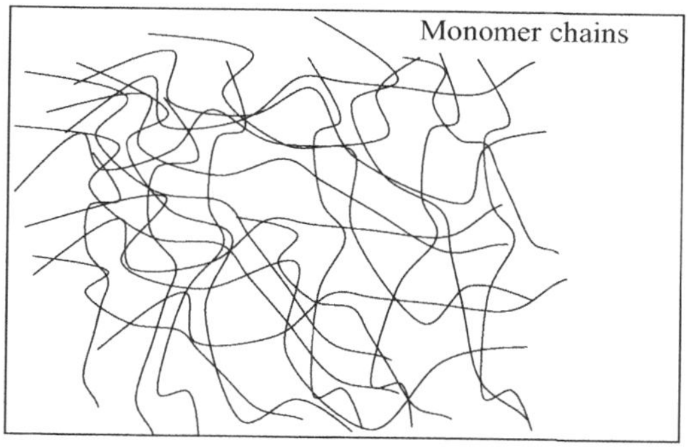

Figure 8.14. *Polymer structure in monomer chains*

If, for example, we track Young's modulus according to temperature, we obtain a characteristic curve in three parts allowing us to depict two characteristic temperatures of the material in question: the glass transition temperature T_g and the melting temperature T_m. Below T_g, the material's behavior is conventional as per a solid material; above T_m, its behavior is almost that of a (more or less viscous) fluid; between the two temperatures, we observe a behavior known as rubbery, characterized by very low rigidity and high deformation capacity. Usually polymers cannot be used as a structural material or as the resin of a composite material at temperatures greater than T_g (although some exceptions to the rule exist, for example when this temperature exceeding T_g is only temporary).

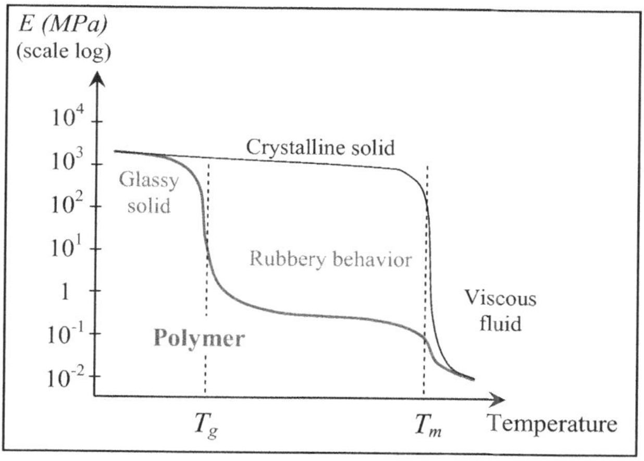

Figure 8.15. *Polymer rigidity according to temperature. For a color version of this figure, see www.iste.co.uk/bouvet/aeronautical.zip*

This rubbery behavior can occur since the molecule chains involved only have weak bonds (hydrogen bonds, Van der Waals bonds, etc.) between one another. This means that they can rearrange themselves, realigning with the loading. This type of polymer is known as a thermoplastic, since its plastic behavior is dependent on temperature.

If we perform a tensile test at a temperature far below its T_g (T_1 in the below figure), we obtain a brittle linear elastic behavior. If we increase the temperature, while staying below the T_g (T_2 in the below figure), we then observe nonlinearity bonded to the viscoelasticity. Viscoelasticity is an intermediate behavior between elasticity, which requires that deformation returns to 0 once the force has been released, and viscous behavior, in which the material flows almost as a liquid.

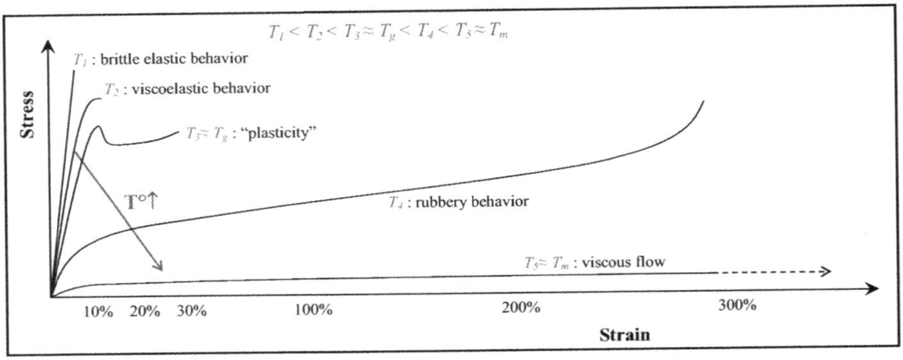

Figure 8.16. *Behavior of polymers with temperature. For a color version of this figure, see www.iste.co.uk/bouvet/aeronautical.zip*

Viscoelastic behavior is highly dependent on stress rate. When the stress is applied quickly, the material reacts immediately, like an elastic material. When the stress is applied slowly or maintained over a long period, the material tends to flow like a viscous fluid.

If we increase the temperature to T_g (T_3 in the figure above), we observe "plasticity" of the material. However, here, the meaning of "plasticity" is taken in the plastic sense of the word, that is, not returning to its initial state once the force is released, and is unrelated to the plasticity of metallic materials. This "plasticity" is due to the molecule chains being stretched by the load, and once stretched, being unable to return to their natural state.

If we increase the temperature beyond T_g (T_4 in the figure above), then we obtain a rubbery elasticity. This elasticity is similar to that of elastomers with a highly nonlinear behavior (the curve begins by softening, then proceeds to harden), but always with a return to zero (or almost zero) deformation on release from stress.

If we increase the temperature still further to around the melting temperature T_m, then we obtain the behavior of a viscous fluid with a nonlinear and perfectly plastic behavior: when the stress is released, the material remains in the deformed state without elastic return.

If we wish to increase the mechanical characteristics of a polymer, we must stop the relative movement of the molecule chains by creating covalent bonds between molecule chains. This phenomenon is called cross-linking.

This is referred to as a thermoset polymer (or a thermoset, the widely used abbreviation, which we will adopt here) if the degree of cross-linking is significant, or as a thermoplastic polymer if cross-linking is absent. It should be noted that the term "thermoset" stems from the fact that the chemical reaction that allows cross-linking is activated by temperature. While this is true during manufacturing, it is not the case latter on once the polymer is already cross-linked.

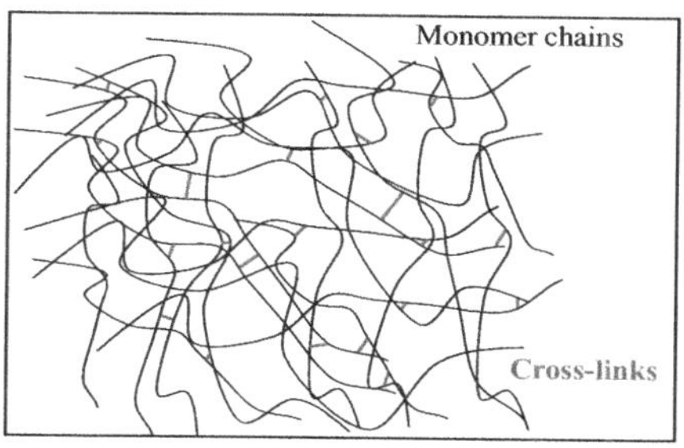

Figure 8.17. *Secondary bonds between monomer chains: cross-links. For a color version of this figure, see www.iste.co.uk/bouvet/aeronautical.zip*

There is a third class of polymer, elastomers, which offer behavior between that of thermosets and thermoplastics. Typically, elastomers have low levels of cross-linking that is characterized by a different sort of covalent bond based on sulfur (referred to as vulcanization). These sulfur-based bonds provide them with very high levels of (largely nonlinear) elasticity.

Now, returning to the change in Young's modulus according to temperature, we are able to differentiate a thermoset polymer from a thermoplastic polymer by a less evident, or absent, glass transition and by the absence of a melting temperature. The glass transition is due to the dissociation of the weak bonds by thermal agitation. Nonetheless, since the thermoset presents numerous cross-links, this cross-link network endures beyond T_g, conferring the thermoset with a good mechanical behavior even after T_g. If we continue to increase temperature, then we reach the material decomposition by pyrolysis, i.e. by decomposition in the form of various gasses and residues (CO, CO_2, H_2, etc.).

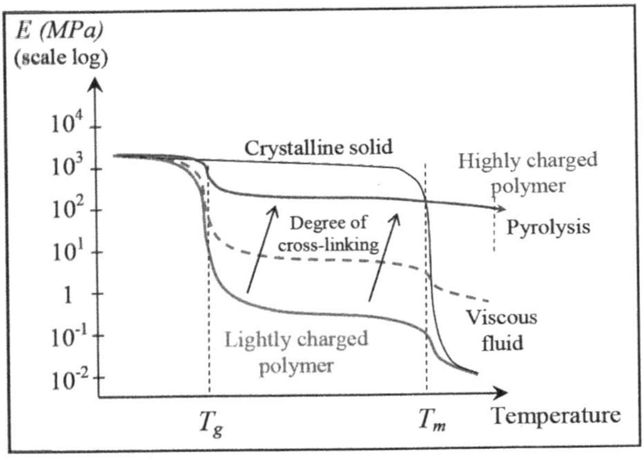

Figure 8.18. *Polymer rigidity according to temperature and degree of cross-linking. For a color version of this figure, see www.iste.co.uk/bouvet/aeronautical.zip*

In chemical terms, this thermoset/thermoplastic difference is due (as a first approximation) to the type of chemical reaction used to synthesize the polymer. Thermoplastic is obtained by a simple addition reaction or polyaddition (making the creation of cross-linking bonds difficult), while a thermoset is obtained by a condensation reaction or polycondensation (inducing many cross-links). This is also known as a polymerization reaction that regroups both of these reactions.

The presence (or absence) of cross-links in the polymer network completely modifies not only its mechanical behavior but also the manufacturing and recycling process. Thermoset is created once and for all during the polycondensation reaction. Accordingly, it must be set in their form prior to this reaction and cannot be reshaped afterwards. Let us consider the example of the carbon/epoxy composite T300/914 mentioned above (epoxy is a thermoset polymer). It is typically sold in the form of film that is 0.1–0.2 mm thick, comprising approximately 50% epoxy matrix and 50% carbon fiber (fiber diameter approximately 7 µm), called prepreg. The name "prepreg" stems from the fact that the fibers are pre-impregnated by resin. This resin is composed of monomers and a hardener, working to favor the chemical reaction of polymerization. In order to avoid the polymerization of prepreg, it has to be maintained at low temperature (typically in the freezer at −20°C) and has a rather short expiration date (typically in the order of 1–2 years). The prepreg must then be assembled in order to create the part required, before the whole is heated. This heating initially acts to cause the resin to melt (or, more precisely, to decrease its viscosity), and thereby give the part its shape. This is a particularly delicate phase since any air bubbles present in the prepreg have to be evacuated, especially those in the layers of the prepreg, in order to avoid any porosity (which act as potential cracks, making the material more brittle). In order to aid the evacuation of these air bubbles, we typically place the thermosetting composite in vacuum sheet, in which we create a primary vacuum, with the whole then being put under pressure (between 5 and 12 bars) and cure. For this, we use an autoclave (essentially, a large pressure cooker), which of course has to be bigger than the workpiece in question; however, this can prove problematic if we need to make the wing of a plane.

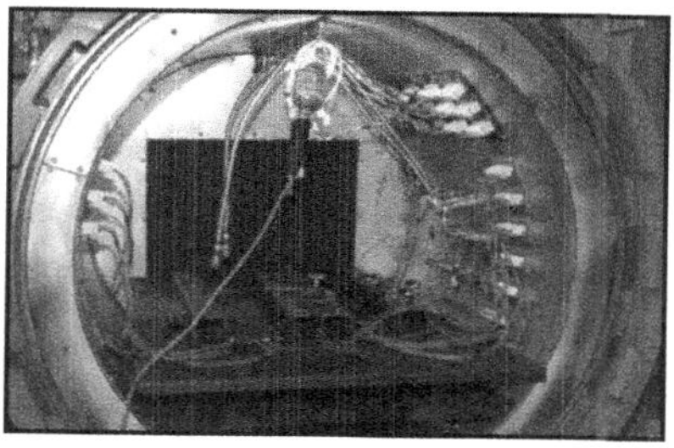

Figure 8.19. *Autoclave for thermoset curing. For a color version of this figure, see www.iste.co.uk/bouvet/aeronautical.zip*

Next, the temperature activates the polymerization reaction (here, polycondensation) and consolidates the material. If we heat the composite a second time then nothing more will happen unless we exceed the pyrolysis temperature. Accordingly, it is impossible to shape a material twice or even to recycle it. Indeed, the need to maintain these materials at low temperature, their short expiration date and the lack of any recycling possibilities constitute the major drawbacks of thermoset polymer.

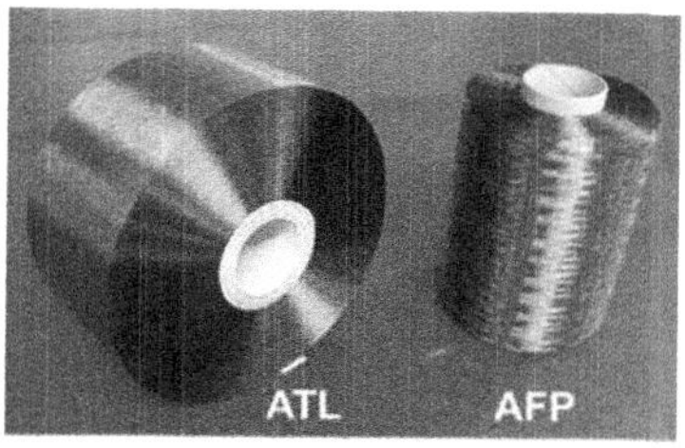

Figure 8.20. *Photo of a prepreg roll (photo Hexcel). For a color version of this figure, see www.iste.co.uk/bouvet/aeronautical.zip*

Unlike thermoset, thermoplastic does not contain cross-linking, which means that it can be reshaped by simply heating it above the melting temperature. As a result, it is easily recyclable and can be kept at ambient temperature with no expiration date. Its major drawback is its lower mechanical characteristics compared to thermoset. Yet despite this, it is currently being tested as a replacement for thermoset in the aeronautical field. It is also undergoing significant scrutiny in the automobile field. It can be shaped by means of Sheet Moulding Compound, or SMC, which allows for high-throughput production. As a comparison, the typical curing cycle of a thermoset composite, such as the T300/914, is around 2 h, as opposed to several minutes for a thermoplastic composite shaped by SMC. Indeed, automobile manufacturers can only consider thermoplastic for production use (outside of a few luxury vehicles).

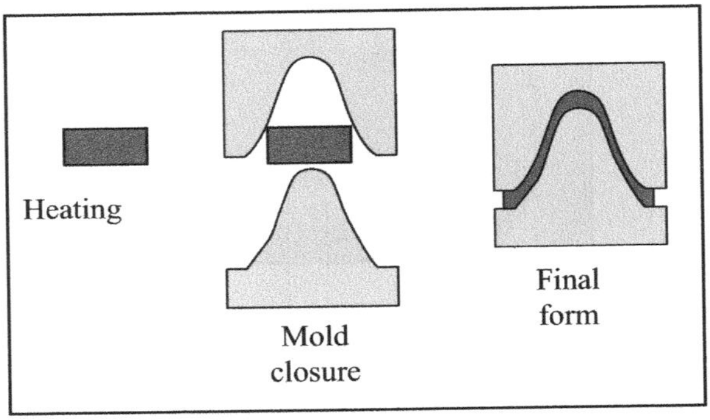

Figure 8.21. *Sheet Moulding Compound (SMC) for thermoplastic heating and shaping. For a color version of this figure, see www.iste.co.uk/bouvet/aeronautical.zip*

Nonetheless, it is possible to increase the mechanical characteristics of a thermoplastic by modifying the molecular chain's morphology. For this, we have to correctly align the molecular chains in order to increase rigidity in the stretching direction. For example, let us consider a tensile test of a thermoplastic polymer. The typical behavior we see here is in three phases:

– the first phase corresponds to the material's linear elasticity;

– if we continue stretching it, we then see a necking (a localized tightening of the cross-section) and plastic behavior (if we release the stress, the deformation does not return to zero). We then see a stress plateau corresponding to the material's stretching and the development of the necking zone throughout the workpiece. Typically, we see maximum strain in the order of 100–300%!

– once the workpiece is stretched, we observe significant rigidification in its mechanical behavior. If we continue to stretch it, we then observe the rupture of the molecular chains being stretched. This is a particularly interesting phase of rigidification since it allows us to obtain a much more rigid material than the initial material.

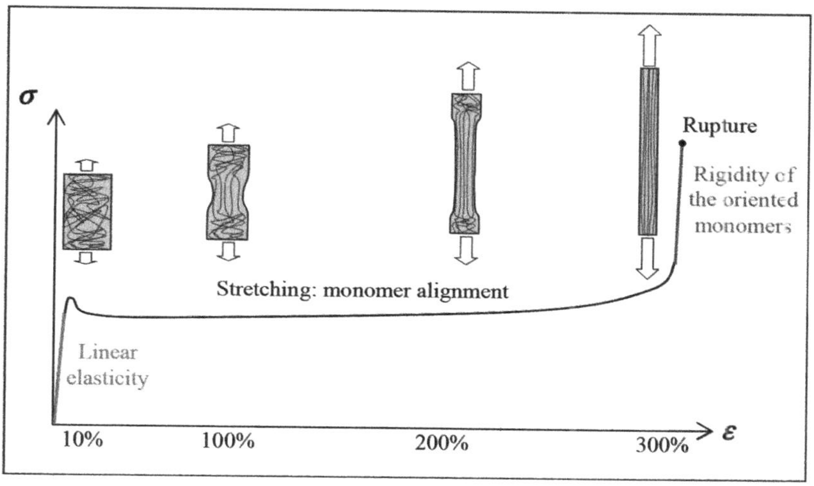

Figure 8.22. *Tension test of a polymer. For a color version of this figure, see www.iste.co.uk/bouvet/aeronautical.zip*

In practice, this stretching allows us to obtain fibers (for example, Nylon fiber is obtained by stretching a liquid form) or thin films, which can be stretched in one or two directions. Of course, this stretching does not provide us with a matrix of composite material, for which we need rigidity in all directions.

Such molecular chain alignments can also be obtained naturally during cooling, which would be crystallization. Indeed, we can show that in order to minimize their energy, molecules attempt to align in certain favored directions. These alignments are only possible in the absence of cross-links, and consequently appear only in thermoplastic.

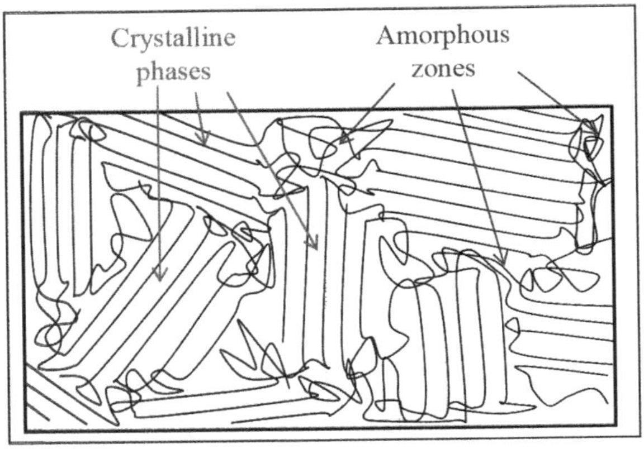

Figure 8.23. *Crystalline phases and amorphous zones in a thermoplastic polymer. For a color version of this figure, see www.iste.co.uk/bouvet/aeronautical.zip*

In practice, the degree of crystallization (or degree of crystallinity), i.e. the portion of the crystalline phase in the material, can reach 80–90% in the most organized thermoplastics, being dependent on the cooling rate. This degree of crystallization is higher as the cooling rate is lower; in other words, the macromolecules must be given time to reorganize if we want crystals. If cooling occurs too quickly, the material forms with no particular order and we obtain an amorphous material; this is a state also known as a glassy or vitreous state. This is also the origin of the term "glass transition", since on passing the glass transition temperature T_g, the crystals disappear and the material again becomes amorphous. It is also due in part to the disappearance of this crystalline phase that thermoplastics' mechanical characteristics decrease significantly on passing T_g.

NOTE.– The glass commonly used in windows is a vitreous material (though by no means a polymer, being largely composed of silica and clearly inorganic). Due to its amorphous state it is transparent, meaning light can pass through it without interacting with the crystalline network (which is absent).

To achieve a thermoplastic matrix with satisfactory mechanical properties, we must not only select an attractive molecule, but also develop a manufacturing process that encourages its crystallization. Now, returning to the development curve of Young's modulus according to temperature, we see that a significantly crystalline thermoplastic behaves in a similar manner to a thermoset with an unremarkable T_g. Of course, we must be mindful not to use this thermoplastic too close to its melting temperature to avoid decreasing its degree of crystallinity and provoke its return to an amorphous state.

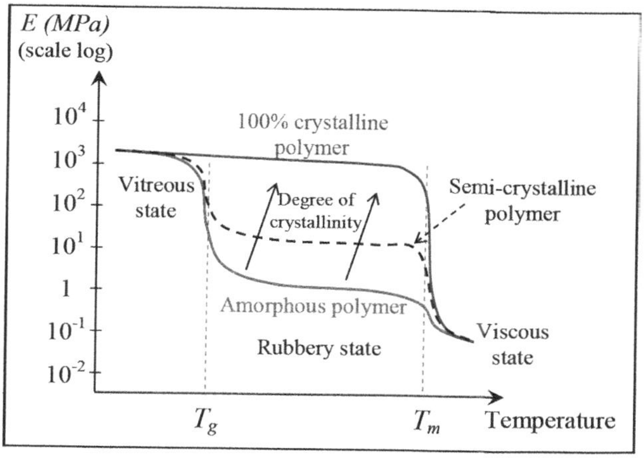

Figure 8.24. *Rigidity of a thermoplastic polymer according to the temperature and degree of crystallinity. For a color version of this figure, see www.iste.co.uk/bouvet/aeronautical.zip*

Currently, the composite materials used for structural applications in aeronautics are in large part composites based on an epoxy matrix; therefore thermoset. Two thermoplastic matrices are under consideration to replace this: the PPS matrix (PolyPhenylene Sulfide) and the PEEK matrix (PolyEther Ether Ketone). The value of thermoplastic is, as was mentioned above, its high recyclability and conservation at ambient temperature with no expiration date. Another advantage is its superior ductility. Since thermoplastic has no cross-links, it is more liable to deform prior to rupture, and presents superior ductility as a consequence. Indeed, remember that one of the major drawbacks of composite material is its brittleness, particularly following impacts, and so matrices with superior ductility let us reduce this brittleness.

In order to compare these three materials, a comparison table is provided below. These values must be taken with precaution, for cost in particular, since this is greatly dependent on the supplier, the quantity produced and the level of the mechanical properties, which depend both on the degree of crystallinity and on temperature. The values provided here correspond to measurements at ambient temperature and a high degree of crystallinity properly controlled during the manufacturing process (in other words, for a material created in good conditions).

	Epoxy matrix (thermoset)	PPS matrix (thermoplastic)	PEEK matrix (thermoplastic)
Density (kg/dm^3)	1.29	1.35	1.32
T_g (°C)	190	90	143
Maximum working temperature (°C)	110	100	260
T_m (°C)	-	285	380
Young's modulus (GPa)	4	3.3	3.3
Tensile stress at rupture (MPa)	100	50	100
Fracture toughness (J/m^2)	100–500	700	4,000
Price (€/kg)	10	10	>100

Table 8.4. *Comparison of three main resins in aeronautics*

Compared with epoxy, PPS has the advantage of a similar cost with somewhat lower mechanical characteristics with the exception of fracture toughness, which is much higher. Its working temperature is also slightly below that of epoxy.

PEEK is a particularly attractive as a candidate to replace epoxy, even if its price still remains prohibitive. It can be used up to 260°C, with mechanical characteristics similar to those of epoxy, with the exception of fracture toughness, which is much higher. Nonetheless, it is more complex to implement, requiring heating above its T_m to be shaped, i.e. temperatures approaching approximately 500°C, while for an epoxy resin a temperature of approximately 190°C is recommended. In all likelihood, we can expect to see its use increase in the coming years once the manufacturing process has been perfected and the cost reduced.

The overview of polymers presented in this paragraph is significantly abridged and the interested reader is referred to [ASH 80], [ASH 86], [DEQ 12], [DOR 86], [DUP 09], [DUP 13], etc.

9

Exercises

9.1. Rosette analysis

A rosette is comprised of three extensometric gages, each of which can measure linear strain relative to its own direction. These instruments, in combination with a measurement chain containing a Wheatstone bridge, are used to determine local strain state through experimentation.

When loaded by a structure with the characteristics E = 70 GPa and ν = 0.3, the gages indicate the following relative elongations:

$$\varepsilon_1 = 1{,}600.10^{-6}\ m/m = 1{,}600\ \mu\varepsilon$$

$$\varepsilon_2 = 800.10^{-6}\ m/m = 800\ \mu\varepsilon$$

$$\varepsilon_3 = -1{,}000.10^{-6}\ m/m = -1{,}000\ \mu\varepsilon$$

Remember that strains have no unit, and since they are small, we often multiply them by 10^6; thereby obtaining the microstrains denoted by με.

The material is assumed to be elastic, linear, homogeneous and isotropic.

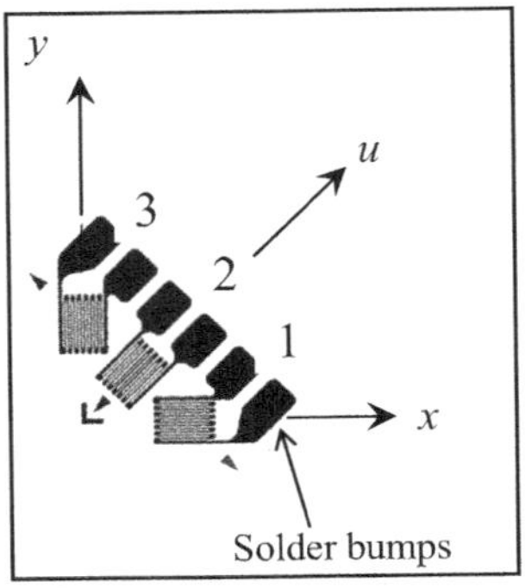

Figure 9.1. *45° strain gage rosette*

QUESTION 1.–

Determine the strain tensor $\underline{\underline{\varepsilon}}(M)$ in 2D.

QUESTION 2.–

Writing that the *(x, y)* plane is a free surface, determine the stress tensor $\underline{\underline{\sigma}}(M)$ in 3D.

QUESTION 3.–

Deduct the strain tensor $\underline{\underline{\varepsilon}}(M)$ in 3D.

QUESTION 4.–

Determine the principal stresses and directions.

QUESTION 5.–

Determine the maximal principal stress together with the associated facet.

QUESTION 6.–

Considering this material as elastic brittle with a tension limit of 100 MPa, will it rupture and in what direction will a crack propagate?

QUESTION 7.–

Determine the maximal shear stress together with the direction of the associated facet.

QUESTION 8.–

Determine the equivalent Von Mises stress.

QUESTION 9.–

Considering this material as ductile with an elastic limit of 140 MPa, will there be plasticity when using a Tresca criterion? And when using a Von Mises criterion?

Further application:

Under the effect of loading on a steel structure ($E = 210\ GPa$, $\nu = 0.3$), a rosette with its three gages at 120° indicates the following relative elongations:

$\varepsilon_1 = -800\ \mu m/m$

$\varepsilon_2 = 1{,}100\ \mu m/m$

$\varepsilon_3 = 1{,}100\ \mu m/m$

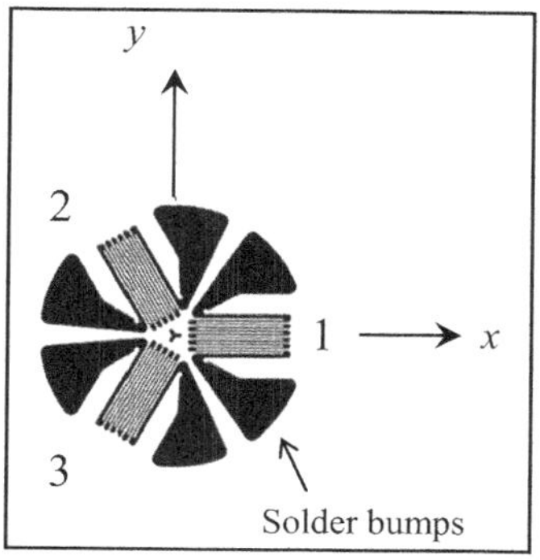

Figure 9.2. *120° strain gage rosette*

Same questions.

9.2. Pure shear

Show that this shear state is a state of pure shear:

$$\underline{\sigma} = \begin{bmatrix} 100 & 100 & 0 \\ 100 & -100 & 0 \\ 0 & 0 & 0 \end{bmatrix} \text{ (in MPa)} \quad [9.1]$$

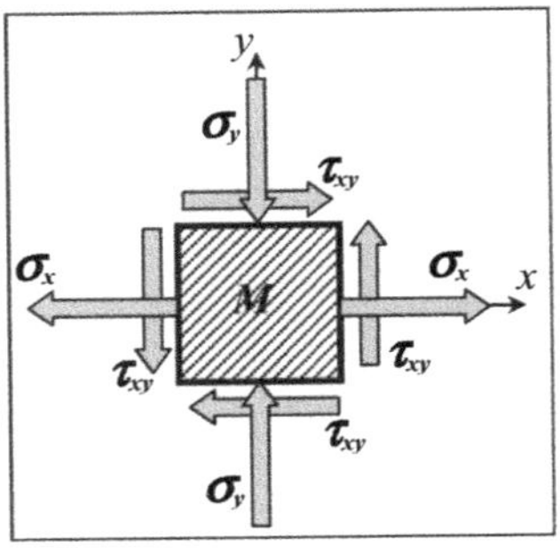

Figure 9.3. *Square sheared!*

Next write the stress tensor in the principal stress coordinate system and in the maximum shear coordinate system.

9.3. Compression of an elastic solid

A material with the characteristics E and ν can be compressed in a cylindrical envelope by means of a piston with cross-section S. Both the envelope and the piston are assumed to be undeformable. The material is assumed to be linear, homogeneous and isotropic.

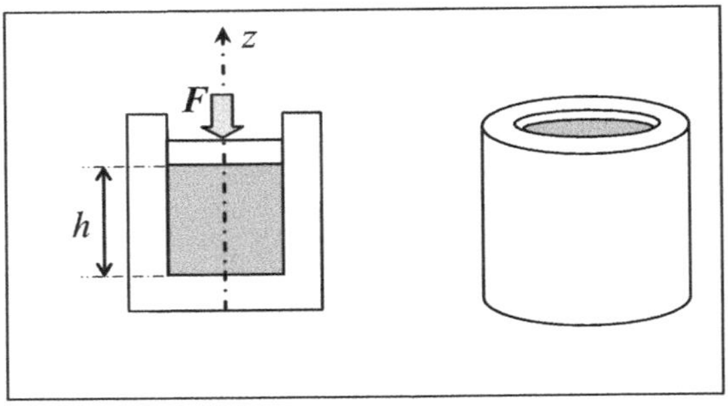

Figure 9.4. *Cylinder confined in compression*

QUESTION 1.–

Determine the strain tensor $\underline{\underline{\varepsilon}}(P)$ and the stress tensor $\underline{\underline{\sigma}}(P)$ at any point P of the compressed material.

QUESTION 2.–

Calculate the piston displacement.

QUESTION 3.–

On what condition is this displacement zero?

9.4. Gravity dam

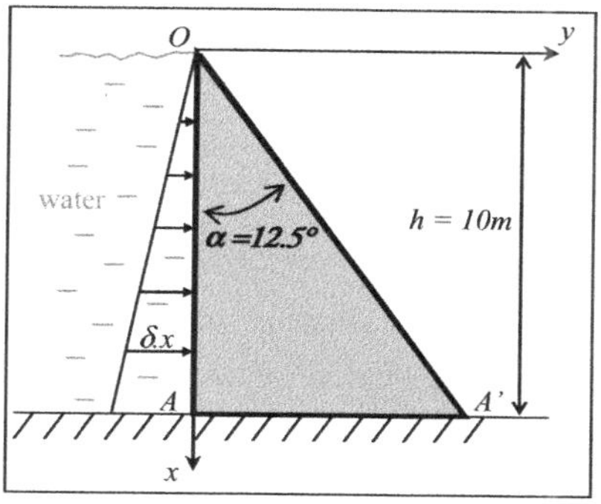

Figure 9.5. *Gravity dam*

For a gravity dam made of concrete:

$$\rho_c = 2{,}500\ kg/m^3,\ E = 2\ GPa,\ \nu = 0.3,\ \sigma^{tens} = 2.5\ MPa,\ \ \sigma^{comp} = -25\ MPa$$

It is delimited by three lines: OA, AA' and OA'. The dam is subjected to its own weight and to forces due to water pressure δx in the y-direction. The forces and stresses in relation to z are assumed to be negligible.

QUESTION 1.–

Determine δ.

QUESTION 2.–

Assuming the stresses to be linear along x and y, determine the stress tensor (in 2D) at all points of the structure.

QUESTION 3.–

At the edges of the dam (for present purposes, and in practice, we can show that it is at the edges that stress is maximum), what are the points at which we find maximum and minimum normal stress and for which directions. Deduce how the dam will break. In practice, how can we reinforce the dam at this point?

9.5. Shear modulus

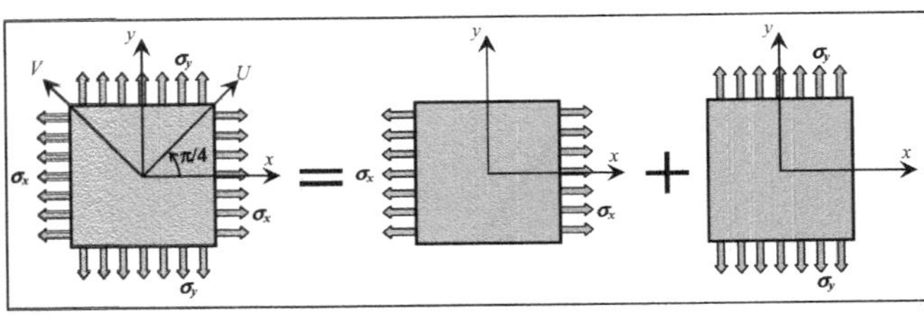

Figure 9.6. *Shear stress analysis*

Here, we will consider a square subjected to a state of stress:

$$\underline{\underline{\sigma}} = \begin{bmatrix} \sigma_x & 0 & 0 \\ 0 & \sigma_y & 0 \\ 0 & 0 & 0 \end{bmatrix} \quad [9.2]$$

Let us assume that the material is linear elastic, homogeneous and isotropic; where E is the Young modulus and ν the Poisson ratio.

QUESTION 1.–

Determine the strain tensor when there is only σ_x.

QUESTION 2.–

Determine the strain tensor when there is only σ_y.

QUESTION 3.–

Using the superposition principle, deduce the strain tensor when there is both σ_x and σ_y.

QUESTION 4.–

Now, we will consider a situation where $\sigma_y = -\sigma_x$.

Determine the strain tensor.

QUESTION 5.–

Determine the stress and strain tensors in the (u, v) basis with $(\underline{x},\underline{u}) = \frac{\pi}{4}$.

Deduce the shear modulus G, defined by $\tau_{xy} = G.\gamma_{xy}$, in relation to E and ν

9.6. Modulus of a composite

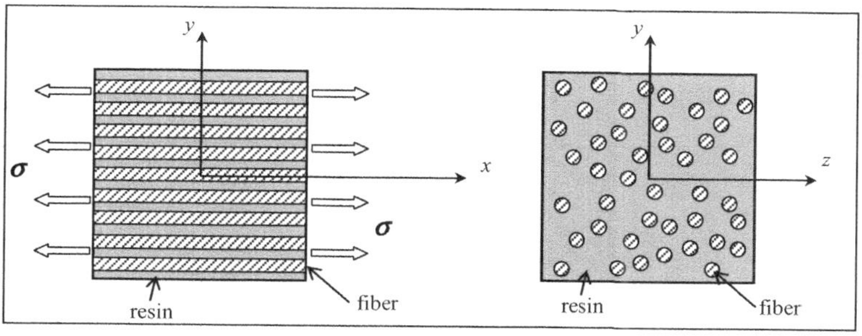

Figure 9.7. *Composite with long fibers*

We will now consider a composite made of carbon fibers ($E_f = 200\ GPa$) in the x-direction and of epoxy resin ($E_r = 10\ GPa$). Here V_f is the volume fraction of fiber and V_r the volume fraction of resin ($V_f + V_r = 1$).

We apply tensile force to this composite in the x-direction.

QUESTION 1.–

Assuming that both materials undergo the same strain, determine the modulus of the composite.

QUESTION 2.–

Assuming that both materials undergo the same stress, determine the modulus of the composite.

QUESTION 3.–

Which of these two hypotheses seems more reasonable? What if we apply tensile force in the y-direction?

QUESTION 4.–

Plot the modulus obtained in relation to the volume fraction of fiber for both of these hypotheses.

NOTE.– The model obtained when assuming homogeneous stresses is called the Voigt limit, while that obtained when assuming homogeneous strains is called the Reuss limit. We can show that the Voigt model is an upper limit and the Reuss model is a lower limit. In practice, we obtain values between them: close to the Voigt limit for traction in the direction of the fibers and close to the Reuss limit for a composite based on particulate reinforcements.

9.7. Torsional cylinder

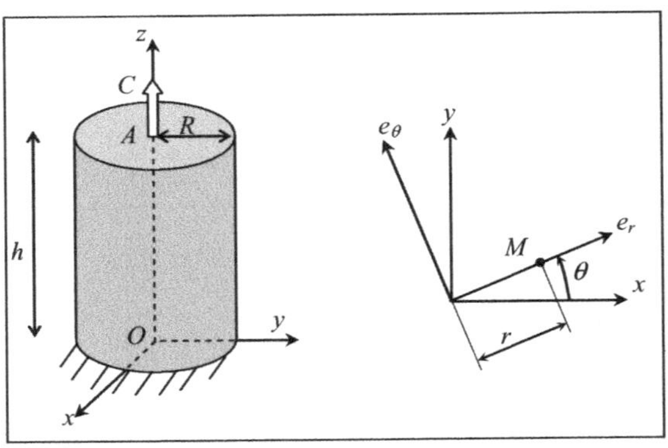

Figure 9.8. *Torsional cylinder*

Here a straight cylinder with circular base of radius R and height h is fixed to the ground. The material is assumed to be linear elastic, homogeneous and isotropic.

We subject it to torque C on the top side, observing movement, such as:

Each straight cross-section turns on its plane around z by a slight angle $d\alpha = k.z$ (k is homogeneous inversely to length with $k << 1/h$)

Cross-section $z = 0$ remains still.

QUESTION 1.–

Find the displacement field of a point $M(r, \theta, z)$.

QUESTION 2.–

Deduce the strain tensor.

QUESTION 3.–

Determine the principal strains and the principal directions.

QUESTION 4.–

Determine the volume expansion. Was the result predicable?

QUESTION 5.–

Determine the stress tensor, and deduce the relationship between k and C.

QUESTION 6.–

Show that the fields of displacement, stress and strain that have been determined are indeed the solution to this problem.

QUESTION 7.–

Assuming that the material is elastic brittle, at what point will it break, for what value of torque C, and in which direction will the crack appear? Test this with a chalk piece.

QUESTION 8.–

Using the same material and retaining the same value of rupture torque, we need to reduce the cylinder's mass. How can we alter its cross-section?

9.8. Plastic compression

We will now consider an aluminum cube of dimension a subjected to compression stress σ in the x-direction. Let us assume that its behavior is perfectly elastic plastic (flow stress is assumed to be constant), with modulus $E = 70\ GPa$ and elastic limit $\sigma_0 = 400\ MPa$.

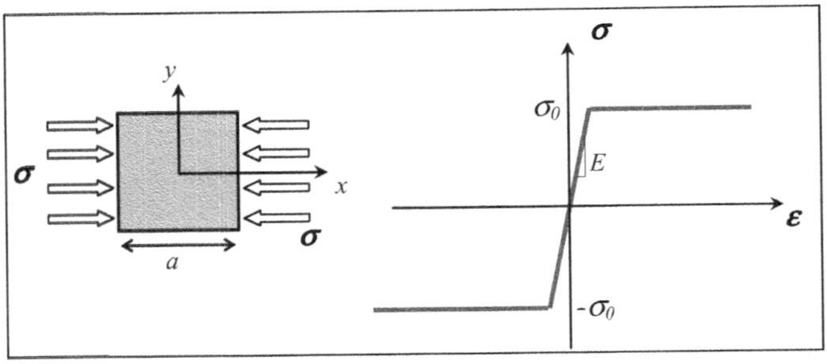

Figure 9.9. *Plastic compression. For a color version of this figure, see www.iste.co.uk/bouvet/aeronautical.zip*

QUESTION 1.–

In practice, we may expect stress to tend to minus infinity when cube thickness tends to 0. Show that this result if coherent with the elastic–plastic behavior proposed, while not confusing true stress/strain with engineering stress/strain.

Deduce the effort/displacement curve obtained in this test.

QUESTION 2.–

In real terms, when performing this test we obtain a barrel effect:

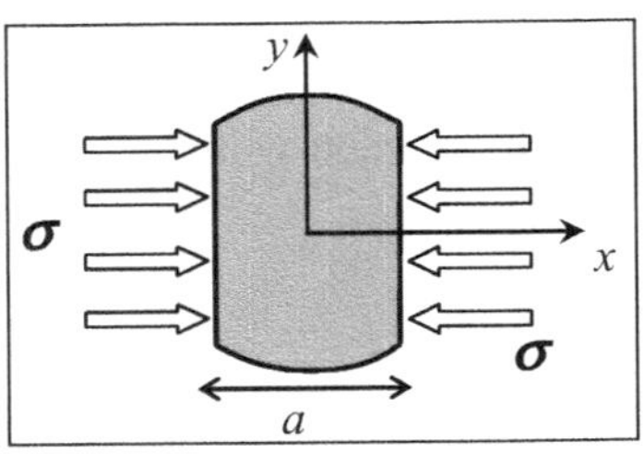

Figure 9.10. *Compressive barrel effect*

Why is this? How do you propose to resolve this problem? And what if we used a longer sample?

QUESTION 3.–

We will now consider a bi-compression test:

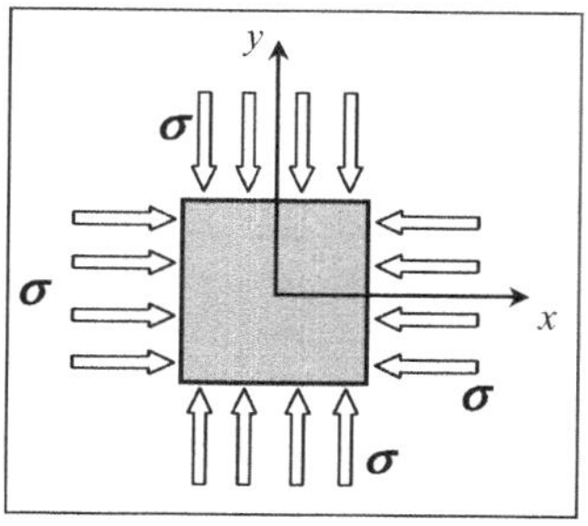

Figure 9.11. *Bi-compression*

At what stress will we obtain plasticity?

QUESTION 4.–

One means of performing this test consists of using beams to press on the cube. In the example shown here, we press on a 2 cm face cube by means of 256 beams with a cross-section of 1 mm^2 and length of 12 mm.

What is the advantage of using these beams?

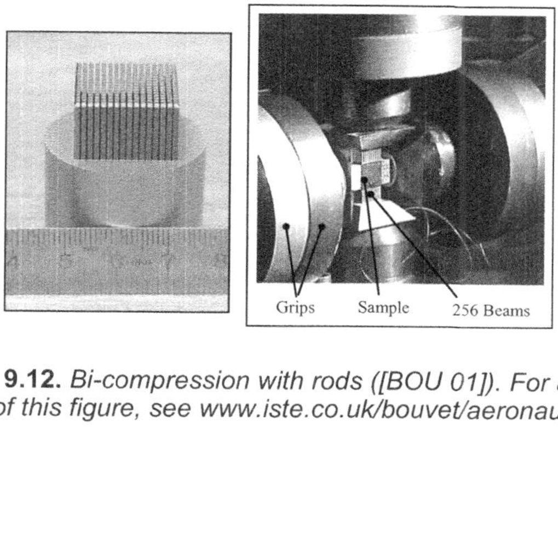

Figure 9.12. *Bi-compression with rods ([BOU 01]). For a color version of this figure, see www.iste.co.uk/bouvet/aeronautical.zip*

QUESTION 5.–

We will now consider a tri-compression test.

At what stress will we obtain plasticity?

9.9. Bi-material beam tension

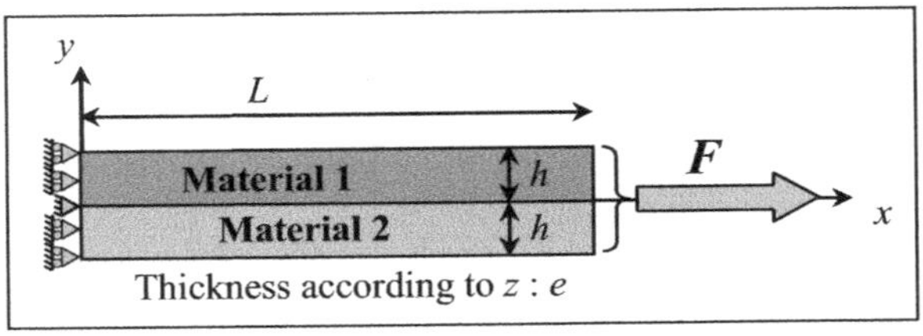

Figure 9.13. *Tension of a bi-material beam*

A beam made of two homogeneous isotropic materials with the respective characteristics E_1, ν_1 and E_2, ν_2 is subjected to tensile effort F.

Part 1: pure tension

Assuming the stress field is in the form:

$$\underline{\sigma}_1 = \begin{bmatrix} \sigma_{x1} & 0 & 0 \\ 0 & 0 & 0 \\ 0 & 0 & 0 \end{bmatrix} \text{ with } \sigma_{x1} = a_1 \text{ in material 1}$$

$$\underline{\sigma}_2 = \begin{bmatrix} \sigma_{x2} & 0 & 0 \\ 0 & 0 & 0 \\ 0 & 0 & 0 \end{bmatrix} \text{ with } \sigma_{x2} = a_2 \text{ in material 2}$$

QUESTION 1.–

Comment on this choice.

QUESTION 2.–

Write out boundary conditions and deduce a relationship between a_1, a_2 and F.

QUESTION 3.–

Determine the strain field.

QUESTION 4.–

Write out the continuity condition at the boundary between the materials 1 and 2 assuming that they are perfectly bonded. Deduce the stress field and plot it.

APPLICATION.– material 1 is steel ($E_1 = 210$ GPa, $\nu_1 = 0.3$) and material 2 aluminum ($E_2 = 70$ GPa, $\nu_2 = 0.3$). To simplify these calculations, take: $E_1 = 3.E_2$ and $\nu_1 = \nu_2$.

QUESTION 5.–

Determine the displacement field.

QUESTION 6.–

Determine the mean apparent modulus of the bi-material.

QUESTION 7.–

Determine the application point of force F. How can we perform the test in reality?

Part 2: tension/bending

Now we need to apply F exactly at the point where $y = 0$:

QUESTION 8.–

Show that if we assume longitudinal strain in the form:

$$\varepsilon_x = a + b.y \qquad [9.3]$$

and stress states such that only σ_x is other than 0 in the two materials, we can then solve the problem. Plot the stress and strain fields in the beam.

9.10. Beam thermal expansion

Part 1: mono-material beam

We will now consider a steel beam, assumed to be linear elastic, homogeneous and isotropic, noted as material 1 below, with characteristics: $E_1 = 210\ GPa$, $\nu_1 = 0.3$, $\alpha_1 = 12.10^{-6}\ K^{-1}$.

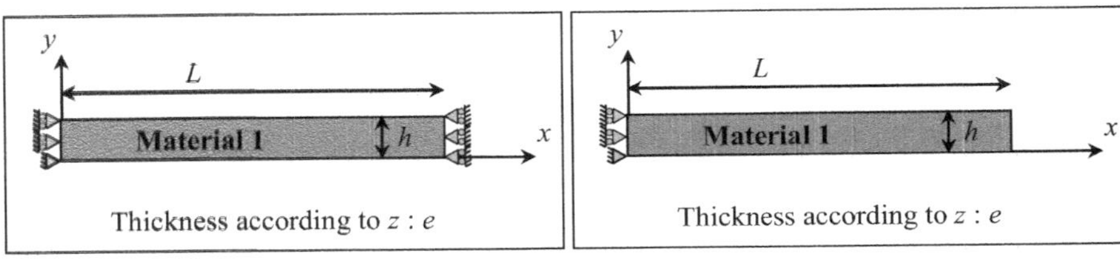

Figure 9.14. *Thermal expansion of a constrained and free beam*

This beam of height $h = 5\ mm$ is at equilibrium at a temperature of $T_0 = 20\ °C$ and is heated to a temperature of $T = 120\ °C$.

QUESTION 1.–

Assuming the beam to be constrained at each end, determine the strain and stress fields.

QUESTION 2.–

Assuming the beam to be free, determine the strain and stress fields.

Part 2: bi-material beam

Now let us consider a bi-material beam made of steel, noted as material 1, and aluminum, noted as material 2, with characteristics:

Steel: $E_1 = 210\ GPa,\ \nu_1 = 0.3,\ \alpha_1 = 12.10^{-6}\ K^{-1}$

Aluminum: $E_2 = 70\ GPa,\ \nu_2 = 0.3,\ \alpha_2 = 24.10^{-6}\ K^{-1}$

These two materials are assumed to be linear elastic, homogeneous and isotropic.

To simplify the calculations, take: $E_1 = 3.E_2$ and $\nu_1 = \nu_2$.

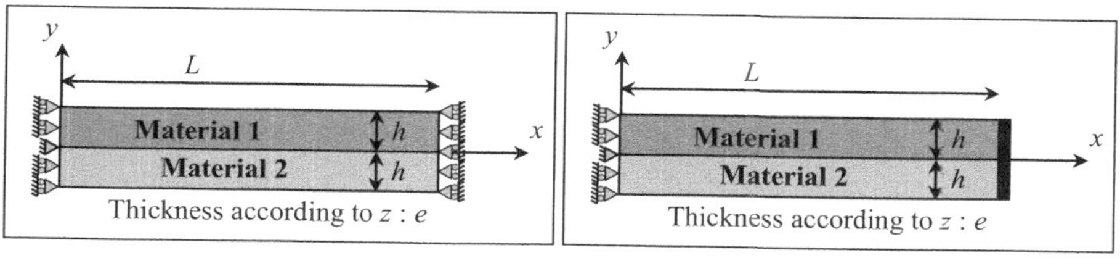

Figure 9.15. *Thermal expansion of a constrained and free bi-material beam*

QUESTION 3.–

Assuming the beam is to be fixed at each end, determine the strain and stress fields in the two materials.

QUESTION 4.–

Assuming the beam is free, determine the strain and stress fields. In this case, we will assume that there is a plate at the end of the beam bonding the two ends on the left and free to the right. Accordingly, the boundary conditions at the end of the two beams are verified along cross-section $x = L$ (rather than at each point).

9.11. Cube under shear stress

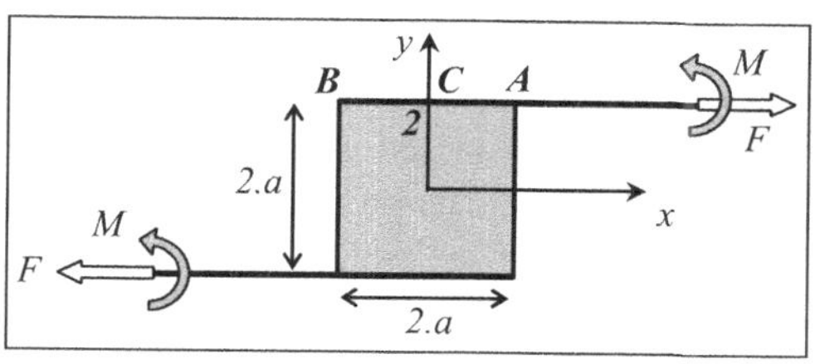

Figure 9.16. *Cube under shear stress*

We will now consider a steel cube subjected to two forces F in the x-direction and two moments M in the z-direction.

QUESTION 1.–

Determine M in relation to F so that the cube is in equilibrium.

We provide: $\sigma_x = 0,\ \sigma_y = 2.K.x.y$ and $\tau_{xy} = K.(a^2 - x^2)$.

QUESTION 2.–

Show that this stress tensor is the solution. Deduce K in relation to F.

QUESTION 3.–

Plot these stress on the four sides of the square.

QUESTION 4.–

Assuming the material to be elastic brittle (for which the maximum stress criteria approximately announce rupture), at what point will it break (if we continue to increase F), and in what direction will the crack appear?

QUESTION 5.–

Assuming the material to be ductile (for which the Tresca criterion approximately announces the end of elasticity), at what point will it begin to become plastic and what will happen if we continue to increase F?

9.12. Spherical reservoir under pressure

The Storable Propellant Stage (*Etage à Propergols Stockables* – EPS) is the upper stage of the Ariane 5, which has the role of adjusting payload orbital insertion in relation to the target orbit and ensuring orientation and separation. Being located inside the launcher, it is not exposed to external conditions. Its design is simple: EPS is a truncated cone-shaped stage pressurized without turbopumps, fitted between the compartment and the payload adapter. Its production is the responsibility of Airbus Defence and Space. The structure weighs 1,200 kg, made up of four aluminum reservoirs containing a total of 9.7 tons of classic MMH (MonoMethyl Hydrazine) propellant and hypergolic liquids (3,200 and 6,500 kg). The structure of EPS is an aluminum honeycomb with the truncated part, a spherical cap supporting the motor, four reservoirs and two helium reservoirs (carbon fiber under 400 bars).

The propellant reservoirs are pressurized by two helium spheres (34 kg) under 400 bars, reduced to 21 bars by a reducer. Without any turbopumps to draw fuel into the reservoirs, it is by means of pressure that the two propellants are injected into the combustion chamber. Outside of the reservoirs is a vacuum.

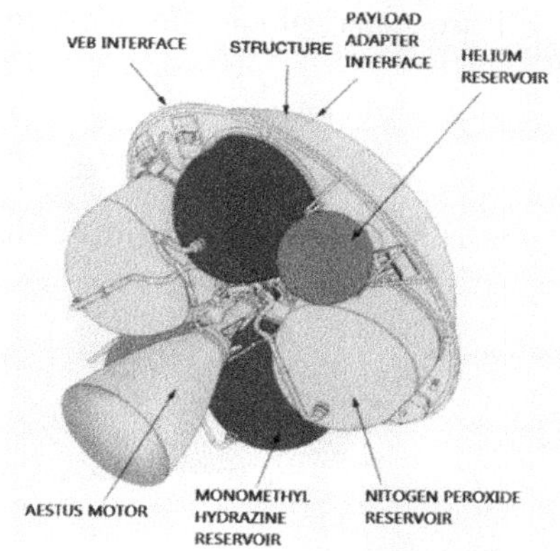

Figure 9.17. *Ariane 5 storable propellant stage (according to http://www.capcomespace.net). For a color version of this figure, see www.iste.co.uk/bouvet/aeronautical.zip*

Here, the objective is to size the helium spheres. Let us assume that these reservoirs are composed of a homogeneous, linear elastic and isotropic material with characteristics E and ν. So, we may define the material's Lamé parameters as:

$$\begin{cases} \lambda = \dfrac{E.\nu}{(1+\nu).(1-2.\nu)} \\ \mu = \dfrac{E}{2.(1+\nu)} = G \end{cases} \quad [9.4]$$

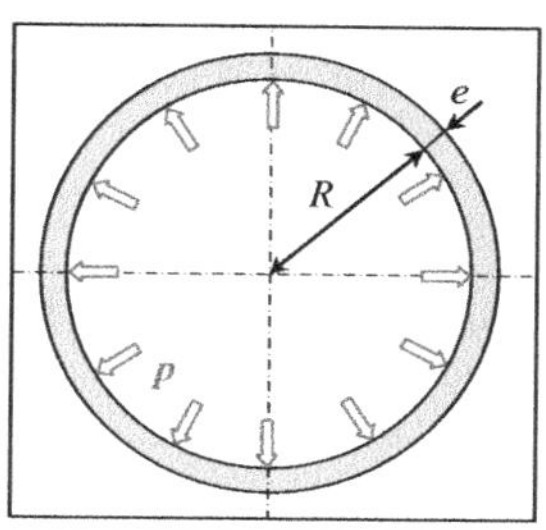

Figure 9.18. *Reservoir under pressure. For a color version of this figure, see www.iste.co.uk/bouvet/aeronautical.zip*

We note $R = 300\ mm$ as the interior radius of the reservoir, e as its thickness and $p = 400\ bars$ its internal pressure. The material is assumed to be linear elastic, homogeneous and isotropic.

In this problem, we will use spherical coordinates: r, θ and ϕ. We will ignore volume forces.

QUESTION 1.–

Let us assume that there is an acceptable kinematic displacement field in the form:

$$\underline{u}(r,\theta,\varphi) = u(r).\underline{e}_r \quad [9.5]$$

Justify this result.

QUESTION 2.–

Specify all of the boundary conditions for this problem.

QUESTION 3.–

What is the Navier's equation? Solve it.

$$\mu.\Delta \underline{u} + (\lambda+\mu).\underline{\underline{\text{grad}}}(\text{div}\underline{u}) + \underline{f} = \underline{0} \quad [9.6]$$

QUESTION 4.–

Express the strain tensor at any point M of the reservoir.

QUESTION 5.–

Express the stress tensor at any point M of the reservoir.

QUESTION 6.–

Determine the integration constants according to the problem's available data, using question 2.

QUESTION 7.–

Does the reservoir volume increase or decrease with deformation?

QUESTION 8.–

We may consider e/R small for the unit. Give an approximation of σ_r, σ_θ and σ_ϕ by performing a first-order limited development.

QUESTION 9.–

We require minimal thickness e for the reservoir, considering that it is made of aluminum with the characteristics:

$$\rho = 2{,}700\ kg/m^3,\ E = 70\ GPa,\ \nu = 0.3,\ \sigma^{el} = 350\ MPa.$$

Deduce reservoir mass.

QUESTION 10.–

We require minimal thickness e for the reservoir, considering that it is made of an epoxy/carbon fabric with the characteristics:

$$\rho = 1{,}800\ kg/m^3,\ E = 70\ GPa,\ \nu = 0.3,\ \sigma^{tens} = 700\ MPa,\ \sigma^{comp} = -400\ MPa$$

Attention, these characteristics are isotropic in the plane, but not isotropic in the out-of-plane directions (in practice, only the resin works in the out-of-plane directions, and we see a modulus of around 10 GPa, a tensile limit around 100 MPa and compression limit around –200 MPa!!)

Deduce the reservoir mass and compare it to that of aluminum.

QUESTION 11.–

Does the hypothesis in question 8 seem justifiable?

QUESTION 12.–

We need to find σ_θ and σ_ϕ using a calculation derived from static of solids. To this end, we will divide the sphere into two hemispheres and apply the fundamental principle of static equilibrium to one of them. Is the result similar to that obtained in question 8?

9.13. *Plastic bending*

Let us now consider an aluminum beam ($E = 70\ GPa$, $\sigma_e = 300\ MPa$) of length L with a rectangular cross-section of height h, width b clamped at $x = 0$ and subjected to a force of F at $x = L$.

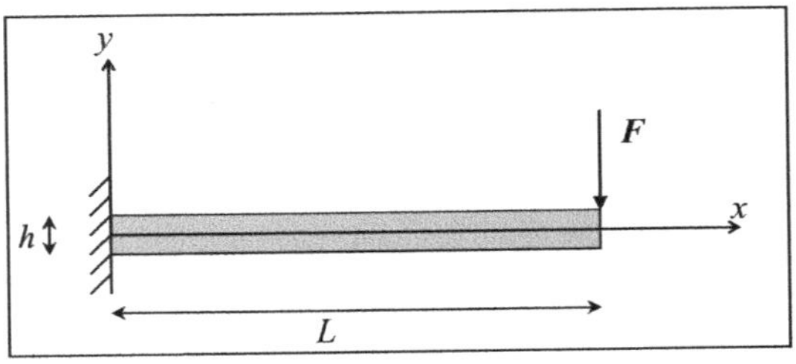

Figure 9.19. *Plastic bending*

QUESTION 1.–

Let us consider the structure in elasticity. Assuming that the longitudinal stress σ_x of a horizontal cross-section x is linear in y:

$$\sigma_x = K(x).y \qquad [9.7]$$

determine $K(x)$ coefficient and show that:

$$K(x) = \frac{-M_{fz}}{I_z} = \frac{12.F.(L-x)}{b.h^3} \qquad [9.8]$$

Plot the stress field σ_x and strain field ε_x in the beam.

QUESTION 2.–

Determine where the plasticity will appear and the value of the elastic effort limit F_e.

QUESTION 3.–

Assuming the material to be perfectly plastic:

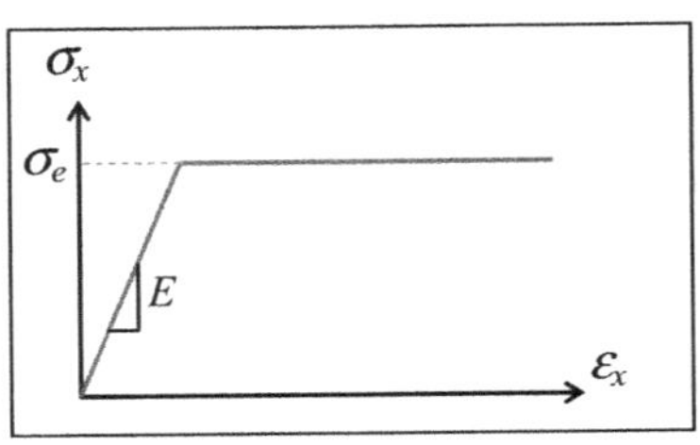

Figure 9.20. *Perfect plasticity. For a color version of this figure, see www.iste.co.uk/bouvet/aeronautical.zip*

Plot the stress field in a cross-section during a test (as F progressively increases). Here, let us assume that the linearity of the strain field ε_x with regard to y remains valid (which would certainly not be the case for that of stress σ_x!).

How will plasticity develop in the beam?

Next, qualitatively plot the development of effort F in relation to the deflection δ, and in particular determine the maximum effort F_{max}.

QUESTION 4.–

What would happen if the effort was released following the development of plasticity? Plot this development on the effort curve in relation to the deflection.

Qualitatively determine the stress field in the cross-section $x = 0$.

9.14. Disc under radial tension

Let us consider a thin steel disc of z-axis (O, z) with interior radius $a = 50\ mm$ and exterior radius $b = 200\ mm$. Thickness e is much less than the other dimensions. We assume the material to be linear elastic, homogeneous and isotropic.

On its periphery, the disc is subjected to a pressure $P = 100\ MPa$ and is clamped on its interior radius.

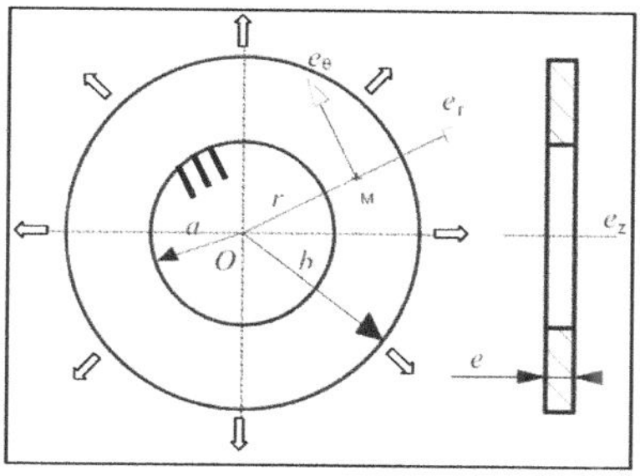

Figure 9.21. *Disc under radial tension*

Part 1: conventional method

QUESTION 1.–

What hypothesis can be associated with this problem? Deduce the form of stress and strain tensors.

QUESTION 2.–

Justify the choice of displacement field:

$$\underline{u}(M) = u(r).\underline{e}_r + w(z).\underline{e}_z \qquad [9.9]$$

Based on the local equilibrium equation expressed in relation to displacement, deduce the differential equation verified by this field.

QUESTION 3.–

Specify the useful boundary conditions to the determination of integration constants. Deduce the expression for radial displacement *u(r)*.

Part 2: energy-based method

QUESTION 4.–

Select an admissible kinematic displacement field (that respects the displacement boundary conditions of the problem) among the linear displacement fields of r.

Discuss the hypothesis of planar stress? Propose a choice of displacement field (but do not use it thereafter) in order to respect the hypothesis of plane stress?

QUESTION 5.–

Calculate the elastic strain energy in the solid.

QUESTION 6.–

Calculate the work of external forces acting on the structure.

QUESTION 7.–

Using the Ritz method, deduce the unknown coefficients of the selected displacement field.

QUESTION 8.–

Compare this approximate solution to the exact solution obtained earlier. In particular, compare the σ_r and σ_θ stresses in $r = a$ and $r = b$. What is the conclusion?

In reality, a significant part of this difference is due to the hypothesis of plane stress. If you redo the first part of this exercise, selecting the hypothesis of plane stain (hypothesis adopted for thick plate), you will find:

$$u(r) = A''.\left(r - \frac{a^2}{r}\right) \qquad [9.10]$$

And:

$$A'' = \frac{P}{2.\left(\lambda + \mu.\left(1 + \frac{a^2}{b^2}\right)\right)} \qquad [9.11]$$

Hence the displacement at $r = b$ is 0.047 mm.

9.15. Bending beam: resolution by the Ritz method

We will now consider a simply supported beam in bending.

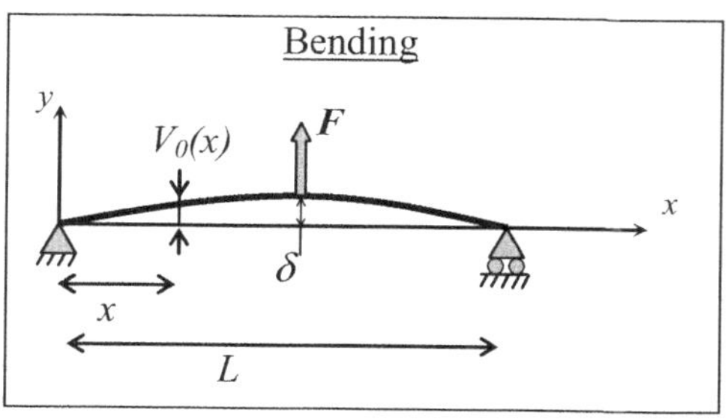

Figure 9.22. *Simply supported beam in bending*

QUESTION 1.–

Propose a deformation shape $v_0(x)$ of the beam neutral axis, with sinusoidal shape, depending only on parameter δ (the beam deflection in center).

QUESTION 2.–

Assuming that a cross-section remains straight and perpendicular to the beam neutral axis, deduce the $u(x, y)$ and $v(x, y)$ displacement fields throughout the beam.

QUESTION 3.–

Deduce the strain field.

QUESTION 4.–

Deduce the stress field, plot it and comment.

QUESTION 5.–

Apply the Ritz method to find the deflection δ in relation to bending force F.

Compare the value of this deflection to the exact solution:

$$f = \frac{F.L^3}{48.E.I_z} \quad [9.12]$$

9.16. Stress concentration in open hole

We will now consider the stress field in a plate drilled by a circular hole of radius R, considered very small relative to the plate dimensions. This plate is submitted to a tensile stress σ_0 in the x-direction "to infinity".

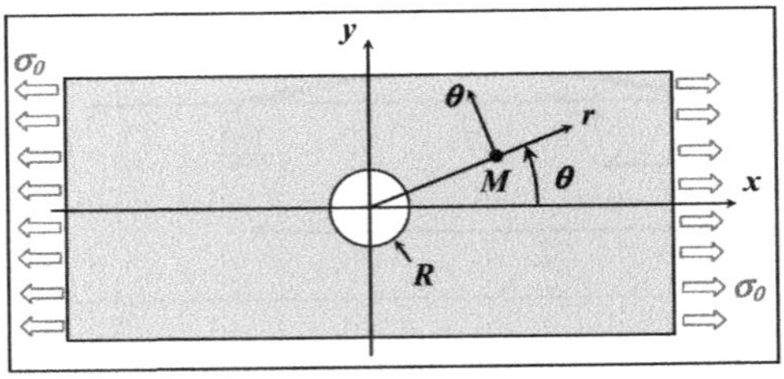

Figure 9.23. *Open hole tension test. For a color version of this figure, see www.iste.co.uk/bouvet/aeronautical.zip*

The plate is of low thickness and we will assume the plane stress:

$$\underline{\underline{\sigma}}(M) = \begin{bmatrix} \sigma_r & \tau_{r\theta} & 0 \\ \tau_{r\theta} & \sigma_\theta & 0 \\ 0 & 0 & 0 \end{bmatrix}_{(\underline{r},\underline{\theta},\underline{z})} \quad [9.13]$$

We can show that the stresses are written as:

$$\begin{cases} \dfrac{\sigma_r}{\sigma_0} = \dfrac{1}{2}.\left[1-\dfrac{1}{\rho^2}+\left(1-\dfrac{4}{\rho^2}+\dfrac{3}{\rho^4}\right).\cos(2.\theta)\right] \\ \dfrac{\sigma_\theta}{\sigma_0} = \dfrac{1}{2}.\left[1+\dfrac{1}{\rho^2}-\left(1+\dfrac{3}{\rho^4}\right).\cos(2.\theta)\right] \\ \dfrac{\tau_{r\theta}}{\sigma_0} = -\dfrac{1}{2}.\left(1+\dfrac{2}{\rho^2}-\dfrac{3}{\rho^4}\right).\sin(2.\theta) \end{cases} \quad \text{with} \quad \rho = \frac{r}{R} \qquad [9.14]$$

Students interested in the proof are referred, for example, to [CHE 08], [AGA 08], [BAM 08], etc. or to the examination that inspired this exercise: "http://mms2.ensmp.fr/mmc_paris/annales/examen2007.pdf".

QUESTION 1.–

Show that this stress field is the solution to the problem.

QUESTION 2.–

Determine and plot the stress field in $r = R$.

QUESTION 3.–

Determine the stress concentration factor at the edge of the hole:

$$K_t = \frac{\sigma_{\max}}{\sigma_0} \qquad [9.15]$$

Considering that the material is brittle with a tension limit stress much less than its compression limit (in absolute value), determine the σ_0 value at which rupture will appear, together with the crack location and the crack propagation direction.

In particular, show that the response is different for tension and compression.

Below are several of the results obtained by finite elements calculation (aluminum material, dimensions 400 × 400 mm^2, hole radius 10 mm, $\sigma_0 = 100\ MPa$). Of course, only a quarter of the structure is meshed, and symmetries are imposed:

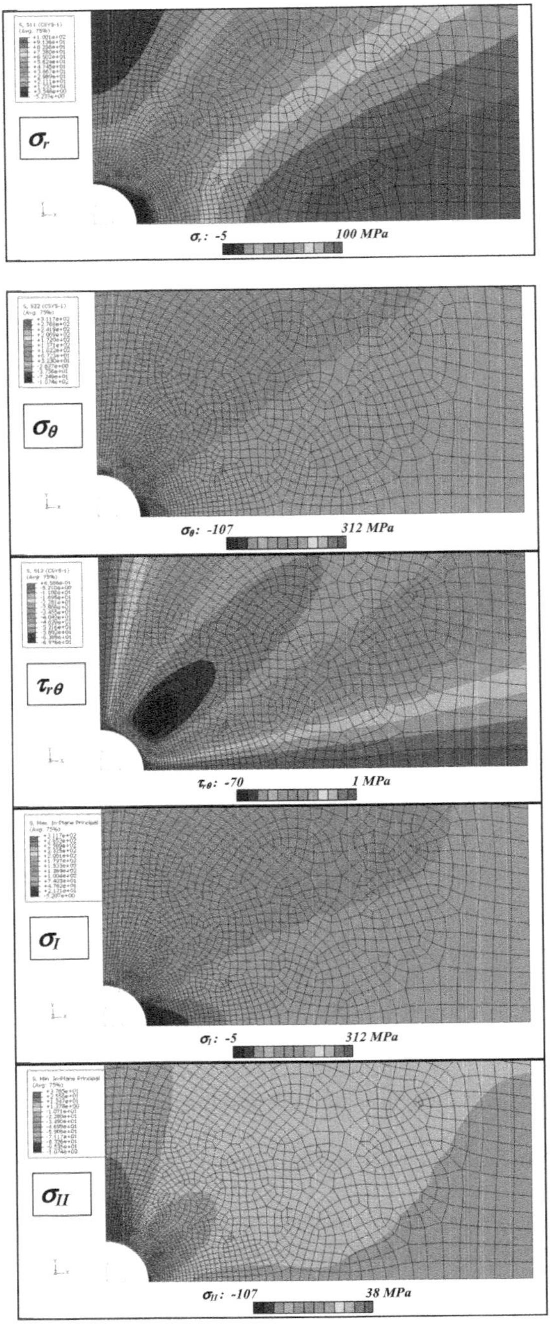
σ_r
σ_r: -5
100 MPa
σ_θ
σ_θ: -107
312 MPa
$\tau_{r\theta}$
$\tau_{r\theta}$: -70
1 MPa
σ_I
σ_I: -5
312 MPa
σ_{II}
σ_{II}: -107
38 MPa

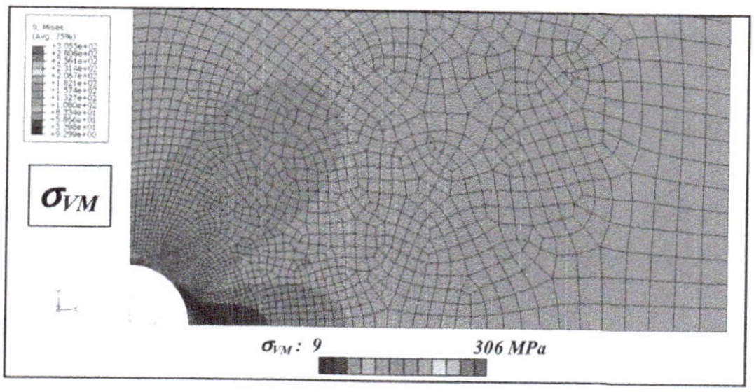

Figure 9.24. *Results of the FE calculation for open hole tension. For a color version of this figure, see www.iste.co.uk/bouvet/aeronautical.zip*

In cases where the hole size is not negligible (dimensions 200 × 40 mm^2), stress concentration increases still further. Essentially, the structure is better without holes. Incidentally, you may also observe that this geometry is representative of a plate drilled with holes spaced by twice hole diameter (in aeronautics, this distance, called pitch distance, is actually four times diameter):

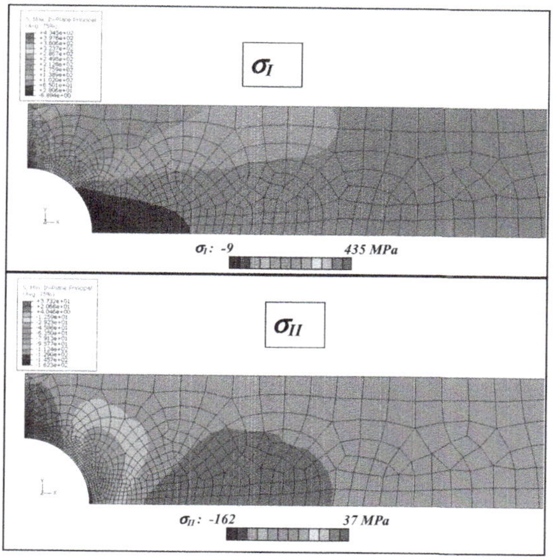

Figure 9.25. *Results of the FE calculation for open hole tension. For a color version of this figure, see www.iste.co.uk/bouvet/aeronautical.zip*

9.17. Bending beam

Part 1: strain tensor

We will now consider a beam of dimension L × h × b bending while clamped at one end and subjected to force F at the other end (face $x = L$).

Let us assume the stress be on the (x, y) plane, solving the problem in 2D.

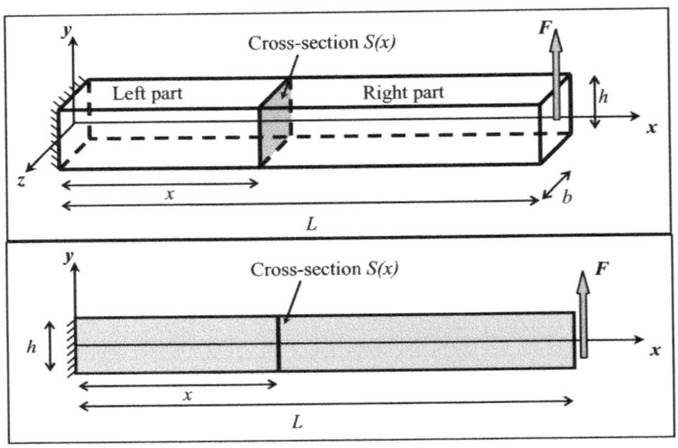

Figure 9.26. *Bending beam*

QUESTION 1.–

Determine the torsor (that is, the force and moment) of internal forces of the left part on the left part of the beam. Show that it can be decomposed into a force in the y-direction, called shear force F_y, and a moment in the z-direction called bending moment M_z. Determine (without calculating them) what stresses are created by F_y and M_z.

QUESTION 2.–

Write out all of the conditions that stress tensor must verify (in 2D) to solve this problem.

QUESTION 3.–

Assuming:

$$\sigma_x = K.\,(L - x).\,y \qquad [9.16]$$

show that we can then solve the problem.

Now solve the problem, determining the stress tensor throughout the beam. In particular, determine K in relation to F and the geometric characteristics of the beam.

Keep this result in relation to K in order to help with writing the following.

QUESTION 4.–

Physically justify the shape of σ_x given by equation [9.16].

QUESTION 5.–

Determine and plot the stress in a horizontal cross-section with an x-coordinate. Deduce the force (force and moment) of these stresses, showing that they are equal to the cohesive forces determined in question 1.

QUESTION 6.–

This beam is made of aluminum with characteristics: $E = 70\ GPa;\ \nu = 0.3;\ \sigma_e = 250\ MPa$

Select a sizing criterion for this beam and justify it. Determine at what points this criterion is maximum (assuming $L > h$).

QUESTION 7.–

Next determine the force F_{lim} for which this criterion is reached. What would happen in reality at such a force, and what would happen if we continued increasing F?

QUESTION 8.–

This beam is made of glass with characteristics: $E = 70\ GPa;\ \nu = 0.3;\ \sigma_r = 60\ MPa$.

Select a sizing criterion for this beam and justify it. Determine at what points this criterion is maximum (assuming $L > h$).

QUESTION 9.–

Next determine the force F_{lim} for which this criterion is reached. What would happen in reality at such a force, and what would happen if we continued increasing F?

Part 2: strain tensor and displacement

QUESTION 10.–

Determine the strain tensor throughout the beam in relation to E and ν.

QUESTION 11.–

Assuming that $\nu = 0$ (to simplify the calculations) and taking the boundary condition as:

$$\begin{cases} u\ (x = 0, y = 0) = 0 \\ v\ (x = 0, y = 0) = 0 \\ u\ (x = 0, y = h/2) = 0 \end{cases} \qquad [9.17]$$

determine the displacement field throughout the beam in relation to K.

NOTE.– We can show that the clamping boundary condition is not coherent with the hypothesis of the stress field found in equation 9.16, but that it induces other stresses approaching the clamping boundary condition.

QUESTION 12.–

Show that the boundary conditions proposed in the previous question let us block any rigid-body displacement field in the plane.

QUESTION 13.–

Deduce the relationship between the beam deflection δ and force F, showing that in the beam case, i.e. where length L is much greater than the other dimensions, we find:

$$f = \frac{F.L^3}{3.E.I_z} \qquad [9.18]$$

with I_z being the quadratic bending moment of inertia in the z-direction:

$$I_z = \frac{b.h^3}{12} \qquad [9.19]$$

Part 3: finite element comparison

A finite element study performed using the Abaqus program and the following dimensions, gave the results below.

QUESTION 14.–

Comment on each figure, comparing it with the analytical solution (for which you will have made the main numerical applications).

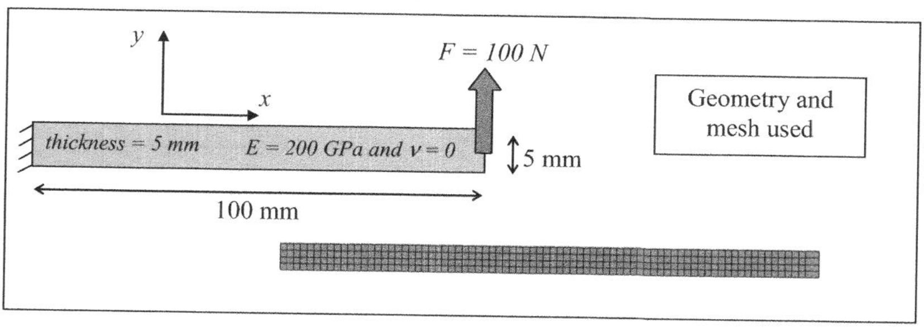

Figure 9.27. *Bending beam. For a color version of this figure, see www.iste.co.uk/bouvet/aeronautical.zip*

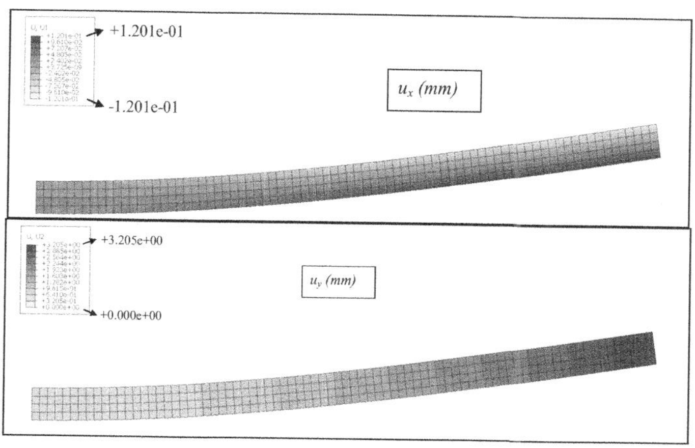

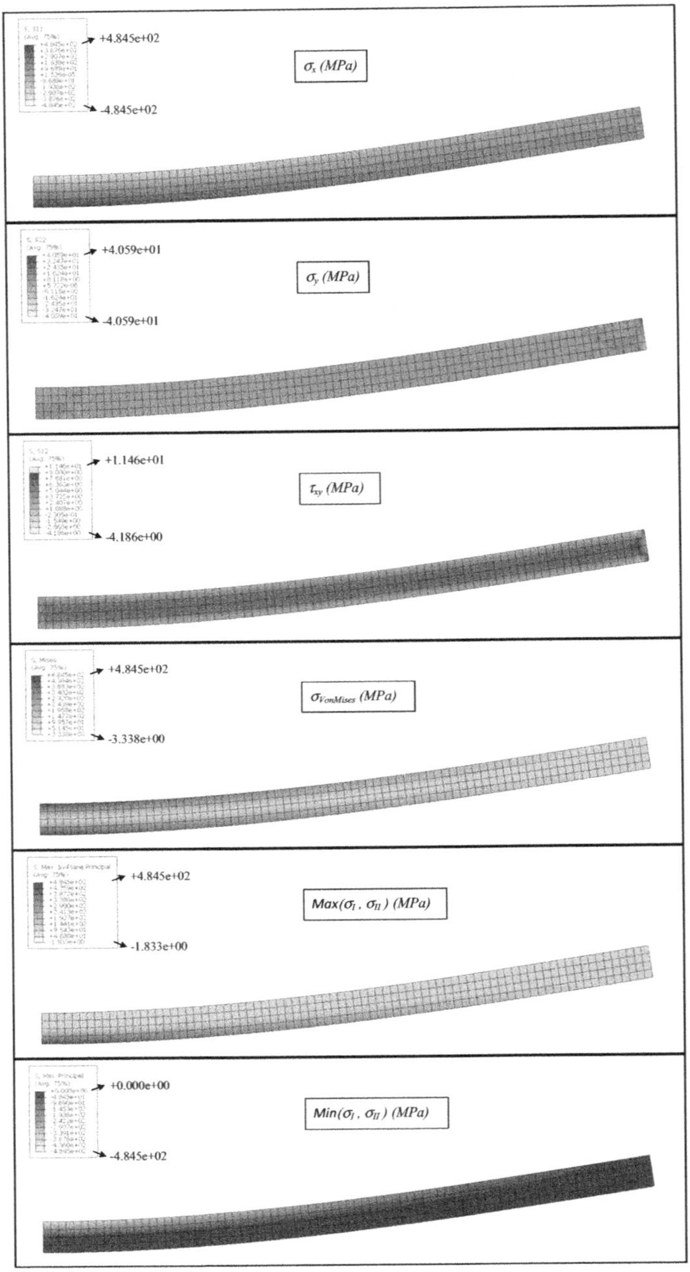

Figure 9.28. *Results of the FE calculation for a bending beam*

10

Solutions to Exercises

10.1. Rosette analysis

QUESTION 1.–

We can simply say that the unit strain in any direction amounts to:

$$\varepsilon(M,\underline{n}) = \varepsilon_n(M) = \underline{n}^t.\underline{\underline{\varepsilon}}(M).\underline{n} \qquad [10.1]$$

Writing this relationship for $\underline{x}$, for $\underline{y}$ and for vector $\underline{u}$ at 45° of $\underline{x}$ (without forgetting to take the unit vector $\underline{u}$, i.e. $\underline{u}$ $(1/\sqrt{2},\ 1/\sqrt{2})$ and removing reference to point M to streamline the notation):

$$\begin{cases} \varepsilon_x = \underline{x}^t.\underline{\underline{\varepsilon}}.\underline{x} = \varepsilon_1 \\ \varepsilon_y = \underline{y}^t.\underline{\underline{\varepsilon}}.\underline{y} = \varepsilon_3 \\ \varepsilon_u = \underline{u}^t.\underline{\underline{\varepsilon}}.\underline{u} = \dfrac{\varepsilon_x + \varepsilon_y}{2} + \varepsilon_{xy} = \varepsilon_2 \end{cases} \qquad [10.2]$$

So, we obtain:

$$\begin{cases} \varepsilon_x = \varepsilon_1 \\ \varepsilon_y = \varepsilon_3 \\ \varepsilon_{xy} = \varepsilon_2 - \dfrac{\varepsilon_1 + \varepsilon_3}{2} \end{cases} \qquad [10.3]$$

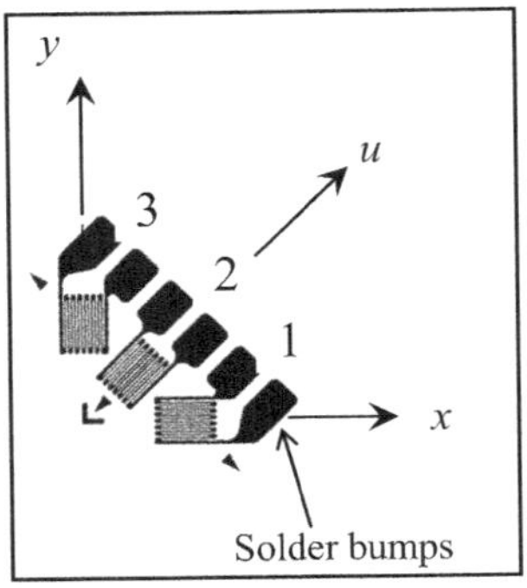

Figure 10.1. *45° strain gage rosette*

That is:

$$\underline{\underline{\varepsilon}}(M)=\begin{bmatrix}\varepsilon_x & \varepsilon_{xy}\\ \varepsilon_{xy} & \varepsilon_y\end{bmatrix}_{(\underline{x},\underline{y})}=\begin{bmatrix}1600 & 500\\ 500 & -1000\end{bmatrix}_{(\underline{x},\underline{y})}\ (in\ \mu\varepsilon) \qquad [10.4]$$

QUESTION 2.–

We have to begin by writing that the external forces applied are zero, and that the external normal direction is equal to $\underline{z}$:

$$\underline{\sigma}(M,\underline{n}_{ext})=\underline{\sigma}(M,\underline{z})=\underline{F}_{ext}=\underline{0} \qquad [10.5]$$

which is:

$$\tau_{xz}=\tau_{yz}=\sigma_z=0 \qquad [10.6]$$

and the stress tensor is consequently plane for on any free surface:

$$\underline{\underline{\sigma}}(M)=\begin{bmatrix}\sigma_{xx} & \tau_{xy} & 0\\ \tau_{xy} & \sigma_{yy} & 0\\ 0 & 0 & 0\end{bmatrix}_{(\underline{x},\underline{y},\underline{z})} \qquad [10.7]$$

We then use the behavior law between stress and strain, assuming the material linear elastic, homogeneous and isotropic:

$$\underline{\underline{\sigma}}=2.\mu.\underline{\underline{\varepsilon}}+\lambda.trace\left(\underline{\underline{\varepsilon}}\right).\underline{\underline{I}} \qquad [10.8]$$

where we can calculate the Lamé parameters using the Young modulus and the Poisson ratio by:

$$\begin{cases} \lambda = \dfrac{E.\nu}{(1+\nu).(1-2.\nu)} \\ \mu = \dfrac{E}{2.(1+\nu)} = G \end{cases} \qquad [10.9]$$

So, we obtain:

$$\begin{cases} \sigma_x = 2.\mu.\varepsilon_x + \lambda.(\varepsilon_x + \varepsilon_y + \varepsilon_z) \\ \sigma_y = 2.\mu.\varepsilon_y + \lambda.(\varepsilon_x + \varepsilon_y + \varepsilon_z) \\ \sigma_z = 2.\mu.\varepsilon_z + \lambda.(\varepsilon_x + \varepsilon_y + \varepsilon_z) = 0 \\ \tau_{xy} = 2.\mu.\varepsilon_{xy} \\ \tau_{yz} = 2.\mu.\varepsilon_{yz} = 0 \\ \tau_{xz} = 2.\mu.\varepsilon_{xz} = 0 \end{cases} \qquad [10.10]$$

Accordingly, the third relationship allows us to determine ε_z:

$$\varepsilon_z = \frac{-\lambda.(\varepsilon_x + \varepsilon_y)}{\lambda + 2.\mu} = -257\ \mu\varepsilon \qquad [10.11]$$

note that we never see plane stress and plane strain at the same time (except when both stress and strain are zero, as may be expected!).

So, we obtain (this is the answer to the next question):

$$\underline{\underline{\varepsilon}}(M) = \begin{bmatrix} 1600 & 500 & 0 \\ 500 & -1000 & 0 \\ 0 & 0 & -257 \end{bmatrix}_{(\underline{x},\underline{y},\underline{z})} \quad (in\ \mu\varepsilon) \qquad [10.12]$$

and

$$\underline{\underline{\sigma}}(M) = \begin{bmatrix} 100 & 27 & 0 \\ 27 & -40 & 0 \\ 0 & 0 & 0 \end{bmatrix}_{(\underline{x},\underline{y},\underline{z})} \quad (in\ MPa) \qquad [10.13]$$

QUESTION 3.–

See above.

QUESTION 4.–

To determine the principal stresses, we must solve:

$$\det\left(\underline{\underline{\sigma}}(M)-\sigma_i.\underline{\underline{I}}\right)=0 \qquad [10.14]$$

This equation admits three solutions, the three principal stresses σ_i *(i = I, II and III)*. We then determine the principal directions by:

$$\begin{cases}\underline{\sigma}(M,\underline{x}_I)=\underline{\underline{\sigma}}(M).\underline{x}_I=\sigma_I.\underline{x}_I\\ \underline{\sigma}(M,\underline{x}_{II})=\underline{\underline{\sigma}}(M).\underline{x}_{II}=\sigma_{II}.\underline{x}_{II}\\ \underline{\sigma}(M,\underline{x}_{III})=\underline{\underline{\sigma}}(M).\underline{x}_{III}=\sigma_{III}.\underline{x}_{III}\end{cases} \qquad [10.15]$$

Here there is evidently an obvious solution:

$$\sigma_{III}=0 \quad \text{and} \quad \underline{x}_{III}=\underline{z}=\begin{bmatrix}0\\0\\1\end{bmatrix} \qquad [10.16]$$

Then we can now work in 2D, which should lead us to:

$$\begin{cases}\sigma_I=105\ MPa\\ \sigma_{II}=-45\ MPa\end{cases} \qquad [10.17]$$

Unless you have inverted the two principal stresses, which of course amounts to the same (the order of the three principal stresses is of no importance and we often class them in ascending order, though this is by no means necessary).

You can now determine $\underline{x}_I$ using the relationship between the principal stress and principal direction:

$$\underline{\sigma}(M,\underline{x}_I)=\underline{\underline{\sigma}}(M).\underline{x}_I=\sigma_I.\underline{x}_I \qquad [10.18]$$

For example, by using $\underline{x}_I(a,b)$ (while still using 2D), we obtain two relationships between a and b, which in fact are the same relationship. Of course, this stems from the fact that $\det\left(\underline{\underline{\sigma}}(M)-\sigma_I.\underline{\underline{I}}\right)=0$. Accordingly, these two coefficients depend on an undetermined multiplicative coefficient; in fact, only the direction of this vector is of interest: its norm is of no importance. For example, we can take:

$$\underline{x}_I = \begin{bmatrix} a \\ b \end{bmatrix} = \begin{bmatrix} 27 \\ 5 \end{bmatrix} \quad [10.19]$$

Here, we may note that this vector is not normalized; it is not difficult to normalize it, but to do so would be pointless!

In fact, this vector $\underline{x}_I$ is clearly 3D, but its coordinate in relation to z is 0 since the problem is plane:

$$\underline{x}_I = \begin{bmatrix} 27 \\ 5 \\ 0 \end{bmatrix} \quad [10.20]$$

We also arrive at this result by recalling that the three principal directions are necessarily orthogonal, and as here, the third principal direction is equal to $\underline{z}$, the other two directions are necessarily in the plane *(x, y)*.

We can now continue with the same reasoning for the last principal direction, or simply use the fact that the principal directions are orthogonal between one another, which here for instance would be (you can easily check that this is indeed the case):

$$\underline{x}_{II} = \begin{bmatrix} -5 \\ 27 \\ 0 \end{bmatrix} \quad [10.21]$$

You can now recall the pattern of principal stresses (in 2D) and verify that the angle between $\underline{x}$ and $\underline{x}_I$ is here 10.5°.

QUESTION 5.–

We can show that the maximum normal stress (for all possible facets) is necessarily the maximum of the three principal stresses, here:

$$\underset{\underline{n}}{Max}(\sigma_n) = Max(\sigma_I, \sigma_{II}, \sigma_{III}) = \sigma_I = 105\ MPa \quad [10.22]$$

And so the direction of the normal vector of this facet is the associated principal direction, here $\underline{x}_I$.

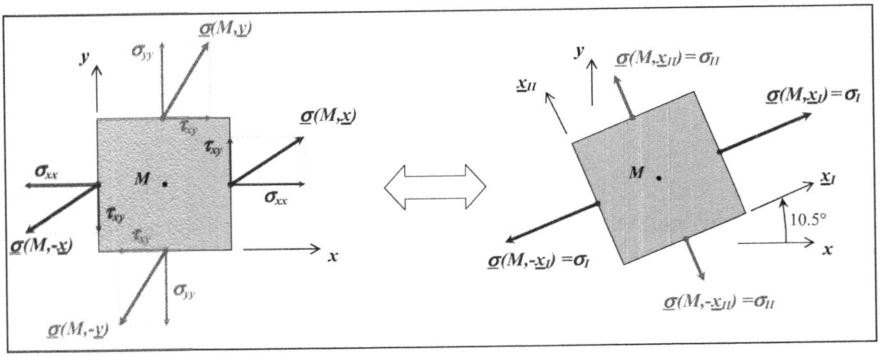

Figure 10.2. *Stress vector and principal stresses. For a color version of this figure, see www.iste.co.uk/bouvet/aeronautical.zip*

QUESTION 6.–

If we consider the material brittle, we can use the normal stress criterion:

$$\underset{\underline{n}}{Max}(\sigma_n) = Max(\sigma_I, \sigma_{II}, \sigma_{III}) < \sigma_{trac} \quad [10.23]$$

which is not verified here, and consequently there would be failure (in practice, stress increases progressively from 0 and failure is reached when the criterion is reached for the first time).

The direction of the crack created by this would be perpendicular to the facet direction of the maximum normal stress, here the facet with normal vector $\underline{x}_I$, and would consequently be in the plane $(\underline{x}_{II}, \underline{x}_{III})$.

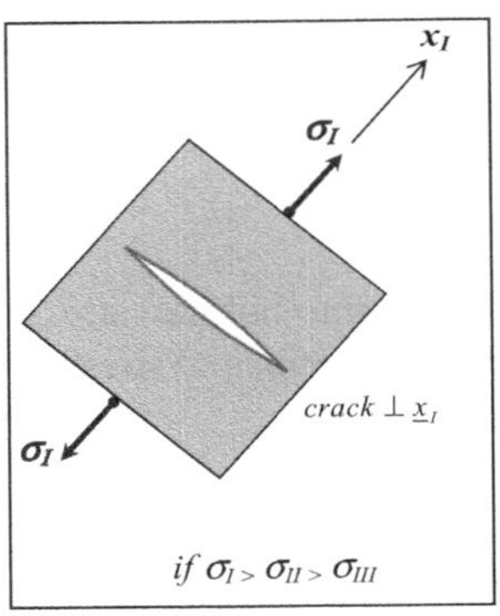

Figure 10.3. *Tension failure of a brittle material. For a color version of this figure, see www.iste.co.uk/bouvet/aeronautical.zip*

QUESTION 7.–

We can show that the maximum shear stress is the maximum between half the difference of the principal stresses:

$$\tau_{\max} = Max\left(\frac{|\sigma_I - \sigma_{II}|}{2};\frac{|\sigma_{II} - \sigma_{III}|}{2};\frac{|\sigma_{III} - \sigma_I|}{2}\right) = \frac{|\sigma_I - \sigma_{II}|}{2} = 75\ MPa \qquad [10.24]$$

And the direction of the associated facet is in the plane $(\underline{x}_I, \underline{x}_{II})$ oriented at 45° of $\underline{x}_I$ or of $\underline{x}_{II}$. Indeed, we can show that the shear stress of a facet of normal vector $(\underline{x}_I + \underline{x}_{II})$ is in direction of $(\underline{x}_{II} - \underline{x}_I)$ and that it remains the same as that of a facet of normal vector $(\underline{x}_{II} - \underline{x}_I)$ (which is in direction of $(\underline{x}_I + \underline{x}_{II})$).

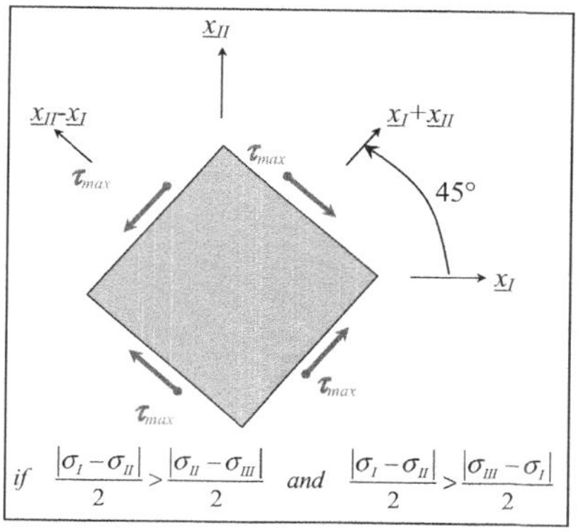

Figure 10.4. *Maximum shear. For a color version of this figure, see www.iste.co.uk/bouvet/aeronautical.zip*

QUESTION 8.–

The Von Mises stress is here:

$$\sigma_{VM} = \sqrt{\frac{1}{2}.\left((\sigma_I - \sigma_{II})^2 + (\sigma_{II} - \sigma_{III})^2 + (\sigma_{III} - \sigma_I)^2\right)} = 133\ MPa \qquad [10.25]$$

QUESTION 9.–

If we consider the Tresca criterion:

$$\sigma_{tresca} = 2.\tau_{\max} = 150\ MPa < \sigma_e \qquad [10.26]$$

This criterion is not verified and consequently there will be plasticity.

Considering the Von Mises criterion:

$$\sigma_{VM} = 133\ MPa < \sigma_e \qquad [10.27]$$

This criterion is verified and so there will be no plasticity.

In conclusion, identifying plasticity is dependent on whether it is the Tresca or Von Mises plasticity criterion that is best adapted to your material.

In practice, since there is little difference between these two criteria, deciding between them can be difficult as the experimental dispersions are often of the same order as the difference between the two criteria.

Further application:

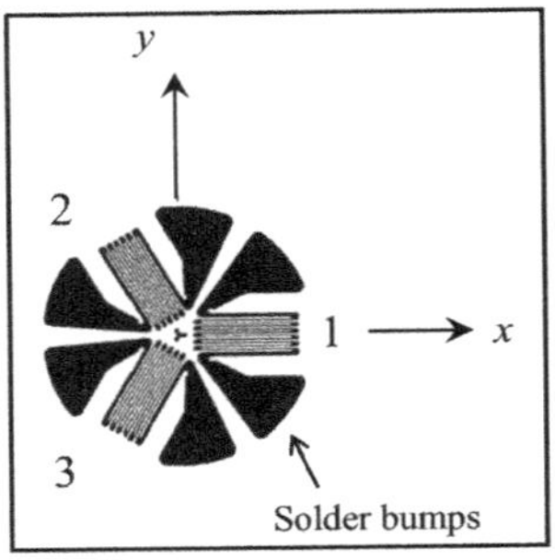

Figure 10.5. *120° strain gage rosette*

We find:

$$\underline{\underline{\varepsilon}}(M) = \begin{bmatrix} -800 & 0 & 0 \\ 0 & 1733 & 0 \\ 0 & 0 & -400 \end{bmatrix}_{(\underline{x},\underline{y},\underline{z})} \quad (in\ \mu\varepsilon) \qquad [10.28]$$

and:

$$\underline{\underline{\sigma}}(M) = \begin{bmatrix} -64.6 & 0 & 0 \\ 0 & 344.6 & 0 \\ 0 & 0 & 0 \end{bmatrix}_{(\underline{x},\underline{y},\underline{z})} \quad (in\ MPa) \qquad [10.29]$$

It is clear then that the principal coordinate system is *(x, y, z)* and the three principal stresses are the three diagonal stresses.

Accordingly, the failure criterion in maximum normal stress:

$$\underset{\underline{n}}{Max}\left(\sigma_n\right)=Max\left(\sigma_I,\sigma_{II},\sigma_{III}\right)=344.6\ MPa<\sigma_{trac} \qquad [10.30]$$

is not verified, so there is failure with a crack in the plane *(x, z)*.

The Tresca criterion:

$$\sigma_{tresca}=2.\tau_{\max}=409.2\ MPa<\sigma_e \qquad [10.31]$$

is not verified; so there is plasticity.

The Von Mises criterion:

$$\sigma_{VM}=381\ MPa<\sigma_e \qquad [10.32]$$

is not verified; so there is plasticity.

10.2. Pure shear

We must simply determine the principal stresses of the stress tensor:

$$\underline{\underline{\sigma}}(M)=\begin{bmatrix}100 & 100 & 0\\ 100 & -100 & 0\\ 0 & 0 & 0\end{bmatrix}_{(\underline{x},\underline{y},\underline{z})}=\begin{bmatrix}100.\sqrt{2} & 0 & 0\\ 0 & -100.\sqrt{2} & 0\\ 0 & 0 & 0\end{bmatrix}_{(\underline{x}_I,\underline{x}_{II},\underline{x}_{III})} \quad (in\ MPa) \qquad [10.33]$$

And observe that in the plane (x_I, x_{II}), by performing a 45° rotation around $\underline{x}_{III}$, we obtain a state of pure shear:

$$\underline{\underline{\sigma}}(M)=\begin{bmatrix}100.\sqrt{2} & 0 & 0\\ 0 & -100.\sqrt{2} & 0\\ 0 & 0 & 0\end{bmatrix}_{(\underline{x}_I,\underline{x}_{II},\underline{x}_{III})}=\begin{bmatrix}0 & -100.\sqrt{2} & 0\\ -100.\sqrt{2} & 0 & 0\\ 0 & 0 & 0\end{bmatrix}_{(\underline{u},\underline{v},\underline{x}_{III})} \quad (in\ MPa) \qquad [10.34]$$

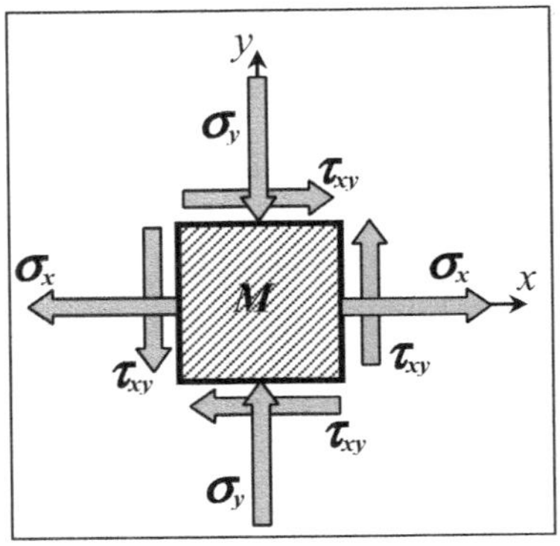

Figure 10.6. *Sheared square*

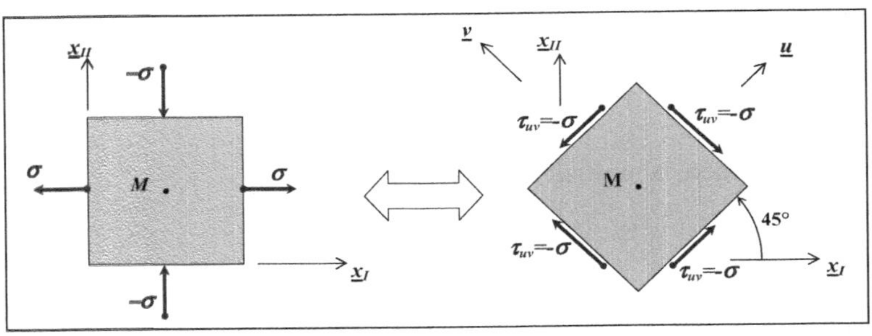

Figure 10.7. *Stress vectors and maximum shear. For a color version of this figure, see www.iste.co.uk/bouvet/aeronautical.zip*

10.3. Compression of an elastic solid

QUESTION 1.–

To find a solution to this problem, we must determine the displacement, stress and strain fields, verifying:

– the equilibrium equation;

– the relationship between displacement and strain;

– the behavior law linking stress and strain;

– the displacement and stress boundary conditions.

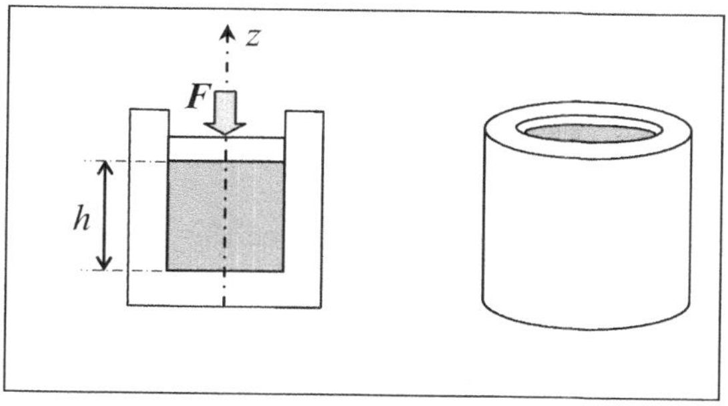

Figure 10.8. *Compression of a confined cylinder*

In addition, as we know that the solution exists and is unique, we just need to find one, and if it meets all of these conditions, it is the right one!

In this case, it seems reasonable to search the displacement field in the form:

$$\underline{u}(P(r,\theta,z)) = w(z).\underline{z} \tag{10.35}$$

In other words, the displacement field is in the z-direction and is function only of z. Again, we do not need to justify this choice, since if it allows us to meet all the conditions then it is the (one and only) solution.

We can then determine the strain tensor:

$$\underline{\underline{\varepsilon}} = \frac{1}{2}.\left(\underline{\underline{grad}}(\underline{u}) + \underline{\underline{grad}}^t(\underline{u})\right) = \begin{bmatrix} 0 & 0 & 0 \\ 0 & 0 & 0 \\ 0 & 0 & w_{,z} \end{bmatrix}_{(\underline{r},\underline{\theta},\underline{z})} \tag{10.36}$$

With the conventional notation:

$$w_{,z} = \frac{\partial w}{\partial z} \tag{10.37}$$

We then determine the stress tensor:

$$\underline{\underline{\sigma}} = 2.\mu.\underline{\underline{\varepsilon}} + \lambda.trace\left(\underline{\underline{\varepsilon}}\right).\underline{\underline{I}} = \begin{bmatrix} \lambda.w_{,z} & 0 & 0 \\ 0 & \lambda.w_{,z} & 0 \\ 0 & 0 & (\lambda+2.\mu).w_{,z} \end{bmatrix}_{(\underline{r},\underline{\theta},\underline{z})} \tag{10.38}$$

The equilibrium equation $\underline{div}\left(\underline{\underline{\sigma}}\right)=\underline{0}$ then leads to:

$$(\lambda+2.\mu).w_{,zz}=0 \quad\Rightarrow\quad w_{,z}=ct=\varepsilon_0 \qquad [10.39]$$

where ε_0 is a constant (equal to strain ε_z, hence this notation). We can then re-determine the stress tensor:

$$\underline{\underline{\sigma}}=\begin{bmatrix}\lambda.\varepsilon_0 & 0 & 0\\ 0 & \lambda.\varepsilon_0 & 0\\ 0 & 0 & (\lambda+2.\mu).\varepsilon_0\end{bmatrix}_{(\underline{r},\underline{\theta},\underline{z})} \qquad [10.40]$$

We must then verify the stress boundary conditions:

– in $z=h$: $\iint_{S_h}\underline{\sigma}(P,\underline{z}).dS=\iint_{S_h}\underline{\underline{\sigma}}.\underline{z}.dS=-F.\underline{z}$

This equation announces the fact that the sum of stress vectors on the top side in $z=h$ is equal to $-F.\underline{z}$.

Hence, by replacing the stress tensor by its expression:

$$(\lambda+2.\mu).\varepsilon_0.S=-F \quad\Rightarrow\quad \varepsilon_0=\frac{-F}{S.(\lambda+2.\mu)}$$

$$\Rightarrow\quad \underline{\underline{\sigma}}=\begin{bmatrix}\frac{-\nu}{1-\nu}.\frac{F}{S} & 0 & 0\\ 0 & \frac{-\nu}{1-\nu}.\frac{F}{S} & 0\\ 0 & 0 & \frac{-F}{S}\end{bmatrix}_{(\underline{r},\underline{\theta}\underline{z})} \qquad [10.41]$$

And here noting that:

$$\frac{\lambda}{\lambda+2.\mu}=\frac{\nu}{1-\nu} \qquad [10.42]$$

We can now determine the displacement field:

$$w(z)=\varepsilon_0.z+cte \qquad [10.43]$$

The boundary conditions then give:

– in $z = 0$: $\underline{u}(P(r,\theta,z=0)) = \underline{0} \Rightarrow w(z=0)=0 \Rightarrow cte=0$;

– in $r = R$: $\underline{u}(P(r=R,\theta,z)).\underline{e}_r = 0$ which is automatically verified here.

Now that all of these conditions have been verified, this is consequently the one and only solution here.

QUESTION 2.–

Piston displacement is the displacement in $z = h$:

$$\Delta w = w(z=h) = \varepsilon_0.h = \frac{-F.h}{S.(\lambda+2.\mu)} = \frac{-F.(1+\nu).(1-2.\nu).h}{E.S.(1-\nu)} \quad [10.44]$$

QUESTION 3.–

Here, we should note that:

$$-1 < \nu \leq 0.5 \quad [10.45]$$

where $\nu \to -1$ corresponds to the case where the shear modulus tends to infinity:

$$G = \frac{E}{2.(1+\nu)} \xrightarrow[\nu\to -1]{} +\infty \quad [10.46]$$

And $\nu = 0.5$ corresponds to an incompressible material. Indeed, we should recall that in tension:

$$\underline{\underline{\sigma}} = \begin{bmatrix} \sigma & 0 & 0 \\ 0 & 0 & 0 \\ 0 & 0 & 0 \end{bmatrix} \Rightarrow \underline{\underline{\varepsilon}} = \begin{bmatrix} \frac{\sigma}{E} & 0 & 0 \\ 0 & \frac{-\nu.\sigma}{E} & 0 \\ 0 & 0 & \frac{-\nu.\sigma}{E} \end{bmatrix} \quad [10.47]$$

And the volume variation is consequently zero irrespective of the stress value:

$$\frac{\Delta V}{V} = tr\left(\underline{\underline{\varepsilon}}\right) = \frac{(1-2.\nu).\sigma}{E} = 0 \quad [10.48]$$

In this case, we also observe that if $\nu = 0.5$, then the piston displacement will be zero irrespective of the force applied. This is a logical result since if the material is incompressible then its volume will not change during compression and piston displacement will remain null.

In practice, some materials, such as rubber for example, do indeed behave in a quasi-incompressible behavior.

10.4. Gravity dam

QUESTION 1.–

Water pressure increases in a linear manner with depth, expressed by the well-known Bernoulli equation:

$$P + \rho.g.z = cte \tag{10.49}$$

except that here z is noted as x, due to $\underline{x}$ is the vertical axis! At point O, pressure is equal to atmospheric pressure, but this pressure is not taken into account since only pressure greater than this atmospheric pressure acts on the structure. So we have:

$$\delta = \rho.g = 10000 \quad N / m^3 \tag{10.50}$$

where ρ is the volumic mass of the water (and not that of the concrete, which we note as ρ_c). Load $\delta.x$ is indeed in N/m^2 and is a surfacic loading.

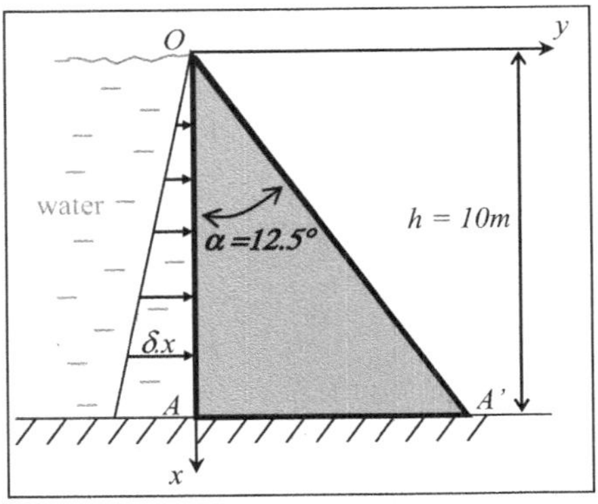

Figure 10.9. *Gravity dam*

QUESTION 2.–

Let us assume the stress tensor is a linear function of x and y:

$$\underline{\underline{\sigma}} = \begin{bmatrix} \sigma_x & \tau_{xy} \\ \tau_{xy} & \sigma_y \end{bmatrix} \quad \text{with} \quad \begin{cases} \sigma_x = a_0 + b_0.x + c_0.y \\ \sigma_y = d_0 + e_0.x + f_0.y \\ \tau_{xy} = g_0 + h_0.x + i_0.y \end{cases} \qquad [10.51]$$

We can begin by asking what justifies this choice. In practice, we see that all loads are linear and all edges of the structure are straight. Of course, this does not prove that the solution is in this form, but since we know that if we indeed find a stress, strain and displacement fields that meet all the conditions, then it is the one and only solution. We must simply make the choice and verify that this choice meets all the conditions, nothing else is needed! The conditions here are:

– the equilibrium equation;

– the relationship between strain and displacement;

– the behavior law linking stress and strain;

– the displacement and stress boundary conditions;

– the compatibility condition.

Indeed, from the moment we make a hypothesis concerning stress (or strain), we must not neglect the compatibility condition. Remember, this condition allows us to both ensure that strain can be integrated and to determine displacement. Essentially, as you perform the hypothesis for the six functions (in 3D, or three in 2D) assuming a shape for the stress (or strain) tensor and that there are only three (in 3D, or two in 2D) for the displacement field, it is by no means clear that everything will be coherent.

The compatibility condition is written as follows:

$$\underline{\underline{grad}}\left(\underline{div}\left(\underline{\underline{\varepsilon}}\right)\right) + \underline{\underline{grad}}^t\left(\underline{div}\left(\underline{\underline{\varepsilon}}\right)\right) - \underline{\underline{\Delta\varepsilon}} - \underline{\underline{grad}}\left(\underline{grad}\left(trace\left(\underline{\underline{\varepsilon}}\right)\right)\right) = \underline{\underline{0}} \qquad [10.52]$$

In practice, we have to determine strain according to stress then check that this equation is indeed verified. Here, this is evident since the compatibility equation only applies for second derivatives. Indeed, since stress is a linear function of x and y, strain will be as well (there is a linear relationship between stress and strain), and all of the secondary derivatives will be zero.

Now we will write out the equilibrium equation:

$$\underline{div}\left(\underline{\underline{\sigma}}\right)+\underline{f}_v=\underline{0} \qquad [10.53]$$

where $\underline{f}_v$ is the volume force due to gravity, that is:

$$\underline{f}_v=\rho_c.g.\underline{x} \qquad [10.54]$$

where ρ_c is the density of concrete. So, now we have $\underline{f}_v$ in N/m^3, which is the force per unit of volume that applies throughout the structure (unlike an external surface force, as in the case of stress boundary conditions which are only applicable on an external surface).

Consequently, the equilibrium equation gives:

$$\begin{cases} b_0+i_0+\rho_c.g=0 \\ h_0+f_0=0 \end{cases} \qquad [10.55]$$

So, we can now rewrite the stress as follows:

$$\begin{cases} \sigma_x=a_0+b_0.x+c_0.y \\ \sigma_y=d_0+e_0.x+f_0.y \\ \tau_{xy}=g_0-f_0.x-(b_0+\rho_c.g).y \end{cases} \qquad [10.56]$$

Now we will write the stress boundary conditions for the side in contact with the water:

$$-\text{ in } y=0:\quad \underline{\sigma}(M,\underline{n}_{ext})=\underline{\sigma}(M,-\underline{y})=\delta.x.\underline{y} \qquad [10.57]$$

hence:

$$\begin{cases} -g_0+f_0.x=0 \\ -d_0-e_0.x=\delta.x \end{cases} \Rightarrow \begin{cases} g_0=f_0=d_0=0 \\ e_0=-\delta \end{cases} \Rightarrow \begin{cases} \sigma_x=a_0+b_0.x+c_0.y \\ \sigma_y=-\delta.x \\ \tau_{xy}=-(b_0+\rho_c.g).y \end{cases} \qquad [10.58]$$

We should note here that for the function $(a.x + b)$ to be zero for any value of x, it must necessarily be the case that a and b are also zero.

Accordingly, the stress boundary conditions for the inclined side in contact with air (note that the equation for this line is $y = x.tan(\alpha)$ and the external normal vector is $\underline{n}_{ext}(-sin(\alpha), cos(\alpha))$):

– in $y = x.tan(\alpha)$: $\underline{\underline{\sigma}}(M, \underline{n}_{ext}) = \underline{0}$

hence:

$$\begin{cases} b_0 + \rho_c.g = \dfrac{\delta}{\tan^2(\alpha)} \\ a_0 = 0 \\ c_0 = \dfrac{\rho_c.g}{\tan(\alpha)} - \dfrac{2.\delta}{\tan^3(\alpha)} \end{cases} \Rightarrow$$

$$\Rightarrow \begin{cases} \sigma_x = \left(\dfrac{\delta}{\tan^2(\alpha)} - \rho_c.g\right).x + \left(\dfrac{\rho_c.g}{\tan(\alpha)} - \dfrac{2.\delta}{\tan^3(\alpha)}\right).y \\ \sigma_y = -\delta.x \\ \tau_{xy} = -\dfrac{\delta}{\tan^2(\alpha)}.y \end{cases} \quad [10.59]$$

Hence the numerical application:

$$\begin{cases} \sigma_x = 178000.x - 1723000.y \\ \sigma_y = -10000.x \\ \tau_{xy} = -203500.y \end{cases} \quad \text{where } \sigma \text{ in Pa and } x, y \text{ in m} \quad [10.60]$$

In practice, we must still show that displacement is zero at the lower part of the dam (in $x = h$). To this end, we have to determine strain, integrate this strain to determine displacement, then verify that it is zero in $x = h$. You can perform these calculations yourselves, but if you complete them, you will see that it does not work! In conclusion, we do not have the exact solution here, and to find it, we have to use a much more complex stress field! However, the solution that we have found here remains a good approximate solution to the real problem.

QUESTION 3.–

To determine whether the dam breaks, we have to use a failure criterion adapted to the dam's concrete. Since concrete is a brittle material, we adapt a criterion of the maximum/minimum principal stress:

$$\begin{cases} Max(\sigma_I, \sigma_{II}, \sigma_{III}) < \sigma_{trac} \\ Min(\sigma_I, \sigma_{II}, \sigma_{III}) > \sigma_{comp} \end{cases} \quad [10.61]$$

Next, we must simply determine the principal stresses at any point of the dam then determine their maximum/minimum values. Since these stresses are linear, by merely looking at the end values, we know that the maximum will be one of these points:

$$\text{at point } O\text{: } \underline{\underline{\sigma}} = \begin{bmatrix} 0 & 0 \\ 0 & 0 \end{bmatrix} \text{ (in MPa), so } \begin{cases} \sigma_I = 0\,MPa \\ \sigma_{II} = 0\,MPa \end{cases};$$

$$\text{at point } A\text{: } \underline{\underline{\sigma}} = \begin{bmatrix} 1.78 & 0 \\ 0 & -0.1 \end{bmatrix} \text{ (in MPa), so } \begin{cases} \sigma_I = 1.78\,MPa \\ \sigma_{II} = -0.1\,MPa \end{cases};$$

$$\text{at point } A'\text{: } \underline{\underline{\sigma}} = \begin{bmatrix} -2.0 & -0.45 \\ -0.45 & -0.1 \end{bmatrix} \text{ (in MPa), so } \begin{cases} \sigma_I = -2.1\,MPa \\ \sigma_{II} = 0.001\,MPa \end{cases}.$$

Accordingly, the failure criterion is verified throughout the structure. It is at point A that we are closest to failure. If water pressure increases (or if the failure criterion of the concrete decreases, due to aging for instance), failure will occur when σ_I reaches the tension failure stress that is 2.5 MPa. Furthermore, the crack will occur perpendicularly to $\underline{x}_I$ (here $\underline{x}_I = \underline{x}$). In practice, we can avoid failure by fitting concrete reinforcing bars in the x-direction at point A; such concrete reinforcing bars would of course be set in a concrete base in the ground.

Here we should note that concrete is a highly asymmetrical material whose compression resistance is around 10 times more than its tension resistance. In other words, we must use concrete mainly for compression and avoid putting it under tension. Typically, we use reinforcing bars in order to alleviate this, or for still greater resistance we can use pre-stressed concrete. For this, iron or steel rods are placed under tension then the concrete is poured around them. Once the concrete is dry, this tensile force on the metal rods is released, which then compresses the concrete. So, even when pre-stressed concrete is put under tension (within a certain threshold) it remains in compression (with its metal rods in tension)!

10.5. Shear modulus

QUESTION 1.–

In order to determine strain from stress, we simply use the behavior law:

$$\underline{\underline{\varepsilon}} = \frac{1+\nu}{E}\underline{\underline{\sigma}} - \frac{\nu}{E}.trace\left(\underline{\underline{\sigma}}\right).\underline{\underline{I}} \tag{10.62}$$

Hence the classic result under tension:

$$\underline{\underline{\varepsilon}} = \begin{bmatrix} \frac{\sigma_x}{E} & 0 & 0 \\ 0 & \frac{-\nu.\sigma_x}{E} & 0 \\ 0 & 0 & \frac{-\nu.\sigma_x}{E} \end{bmatrix}_{(\underline{x},\underline{y},\underline{z})} \tag{10.63}$$

So, the material is elongated according to the tension x-direction and shortens in the two perpendicular directions.

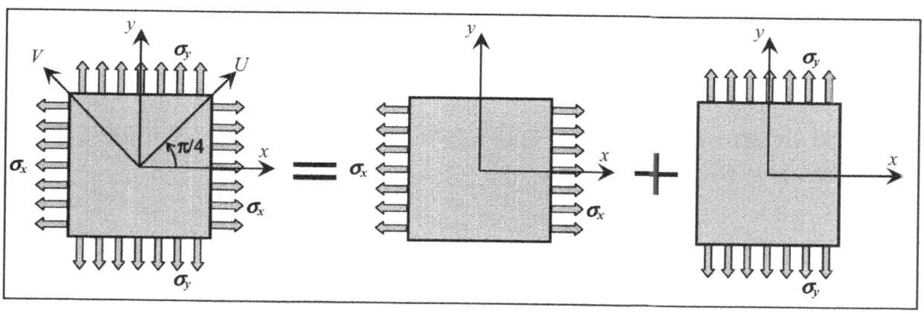

Figure 10.10. *Shear stress analysis*

QUESTION 2.–

As above:

$$\underline{\underline{\varepsilon}} = \begin{bmatrix} \frac{-\nu.\sigma_y}{E} & 0 & 0 \\ 0 & \frac{\sigma_y}{E} & 0 \\ 0 & 0 & \frac{-\nu.\sigma_y}{E} \end{bmatrix}_{(\underline{x},\underline{y},\underline{z})} \tag{10.64}$$

QUESTION 3.–

We simply use the strain from questions 1 and 2:

$$\underline{\underline{\varepsilon}} = \begin{bmatrix} \frac{\sigma_x - \nu.\sigma_y}{E} & 0 & 0 \\ 0 & \frac{\sigma_y - \nu.\sigma_x}{E} & 0 \\ 0 & 0 & \frac{-\nu.(\sigma_x + \sigma_y)}{E} \end{bmatrix}_{(\underline{x},\underline{y},\underline{z})} \quad [10.65]$$

QUESTION 4.–

If $\sigma_y = -\sigma_x$, then:

$$\underline{\underline{\varepsilon}} = \frac{\sigma_x}{E} . \begin{bmatrix} 1+\nu & 0 & 0 \\ 0 & -1-\nu & 0 \\ 0 & 0 & 0 \end{bmatrix}_{(\underline{x},\underline{y},\underline{z})} \quad [10.66]$$

QUESTION 5.–

In order to determine the stress and strain tensor in *(u, v)*, we simply perform a 45° rotation relative to $\underline{z}$.

$$\underline{\underline{\sigma}}_{(\underline{u},\underline{v},\underline{z})} = \underline{\underline{P}}^t . \underline{\underline{\sigma}}_{(\underline{x},\underline{y},\underline{z})} . \underline{\underline{P}} \quad [10.67]$$

where $\underline{\underline{P}}$ is the rotation matrix (and here, of course, with $\theta = 45°$):

$$\underline{\underline{P}} = \begin{bmatrix} \cos(\theta) & -\sin(\theta) & 0 \\ \sin(\theta) & \cos(\theta) & 0 \\ 0 & 0 & 1 \end{bmatrix}_{(\underline{x},\underline{y},\underline{z})} = \begin{bmatrix} 1/\sqrt{2} & -1/\sqrt{2} & 0 \\ 1/\sqrt{2} & 1/\sqrt{2} & 0 \\ 0 & 0 & 1 \end{bmatrix}_{(\underline{x},\underline{y},\underline{z})} \quad [10.68]$$

Hence:

$$\underline{\underline{\sigma}} = \begin{bmatrix} \sigma_x & 0 & 0 \\ 0 & -\sigma_x & 0 \\ 0 & 0 & 0 \end{bmatrix}_{(\underline{x},\underline{y},\underline{z})} = \begin{bmatrix} 0 & -\sigma_x & 0 \\ -\sigma_x & 0 & 0 \\ 0 & 0 & 0 \end{bmatrix}_{(\underline{u},\underline{v},\underline{z})} \quad [10.69]$$

Conventional result illustrated by:

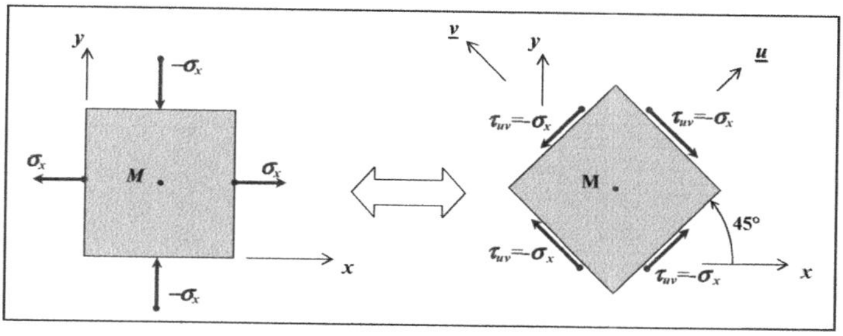

Figure 10.11. *Stress vectors and maximum shear. For a color version of this figure, see www.iste.co.uk/bouvet/aeronautical.zipEssentially, in the coordinate system (u,v,z) we have:*

Again as for stress, we perform a rotation of the strain tensor:

$$\underline{\underline{\varepsilon}} = \frac{\sigma_x.(1+\nu)}{E} . \begin{bmatrix} 1 & 0 & 0 \\ 0 & -1 & 0 \\ 0 & 0 & 0 \end{bmatrix}_{(\underline{x},\underline{y},\underline{z})} = \frac{\sigma_x.(1+\nu)}{E} . \begin{bmatrix} 0 & -1 & 0 \\ -1 & 0 & 0 \\ 0 & 0 & 0 \end{bmatrix}_{(\underline{u},\underline{v},\underline{z})} \quad [10.70]$$

Essentially, in the coordinate system (u, v, z) we have:

$$\begin{cases} \tau_{uv} = -\sigma_x \\ \gamma_{uv} = 2.\varepsilon_{uv} = -\dfrac{2.(1+\nu)}{E} . \sigma_x \end{cases} \quad [10.71]$$

And by definition of the shear modulus G, we have:

$$\tau_{uv} = G.\gamma_{uv} \quad [10.72]$$

So, we have:

$$G = \frac{E}{2.(1+\nu)} \quad [10.73]$$

10.6. Modulus of a composite

QUESTION 1.–

If we assume that both materials show the same strain, then we have:

$$\varepsilon_f = \varepsilon_r = \bar{\varepsilon} \quad [10.74]$$

where index f corresponds to fibers, index r to resin and $\bar{\varepsilon}$ to mean strain.

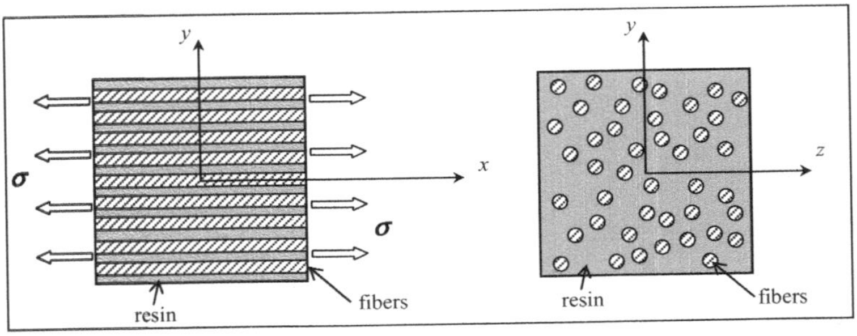

Figure 10.12. *Composite with long fibers*

This amounts to a parallel consideration of the two materials:

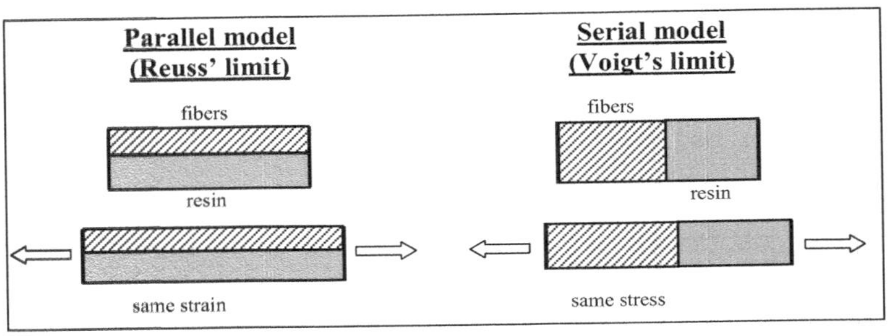

Figure 10.13. *Parallel and serial homogenization models*

Accordingly, the total stress is the sum of the stresses for the two materials (not forgetting to account for the fiber volume fraction V_f; while that of resin is of course $1-V_f$):

$$\bar{\sigma} = V_f.\sigma_f + (1-V_f).\sigma_r \quad [10.75]$$

If we consider an elastic behavior for the two materials:

$$\begin{cases} \sigma_f = E.\varepsilon_f \\ \sigma_r = E.\varepsilon_r \end{cases} \quad [10.76]$$

Then we find that the mean modulus is the mean of the two moduli:

$$\overline{E} = \frac{\overline{\sigma}}{\overline{\varepsilon}} = V_f.E_f + (1 - V_f).E_r \qquad [10.77]$$

QUESTION 2.–

Assuming that the two materials show the same stress, then we have:

$$\sigma_f = \sigma_r = \overline{\sigma} \qquad [10.78]$$

This amounts to a serial consideration of the two materials (see Figure 10.13).

Accordingly, the total strain is the sum of the strains for the two materials (not forgetting to account for the fiber and resin volume fraction):

$$\overline{\varepsilon} = V_f.\varepsilon_f + (1 - V_f).\varepsilon_r \qquad [10.79]$$

Thus, we obtain the mean modulus:

$$\overline{E} = \frac{\overline{\sigma}}{\overline{\varepsilon}} = \frac{1}{\frac{V_f}{E_f} + \frac{1 - V_f}{E_r}} \qquad [10.80]$$

QUESTION 3.–

Which hypothesis seems the more reasonable depends on the direction of tension: if we pull in the fiber direction *(x)*, the parallel model seems better, but if we pull in the transversal direction to the fibers (*y* or *z*), the serial model seems better. Moreover, this is what we observe through experimentation.

QUESTION 4.–

If we plot the mean modulus, then we obtain the results in Figure 10.14.

The modulus that we obtain, assuming homogeneous stress, is called the Voigt limit and the one obtained assuming homogeneous strain is called the Reuss limit. We can show that the Voigt model constitutes an upper bound and the Reuss' model a lower bound. In practice, we obtain values between these two: approaching the Voigt limit for tension in the fiber direction and approaching the Reuss' limit for particulate reinforced composite.

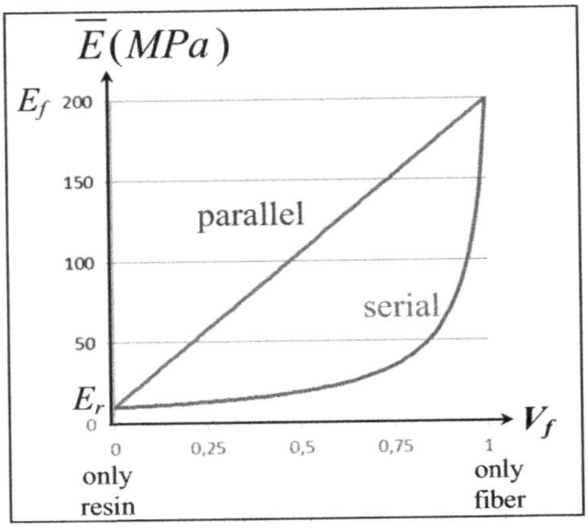

Figure 10.14. *Young's modulus according to parallel and serial homogenization models. For a color version of this figure, see www.iste.co.uk/bouvet/aeronautical.zip*

10.7. Torsional cylinder

QUESTION 1.–

The displacement of point $M(r, \theta, z)$ is equal to the product of the radius by the angle of rotation:

$$\underline{u}(M(r,\theta,z)) = r.d\theta.\underline{e}_\theta = k.r.z.\underline{e}_\theta \quad [10.81]$$

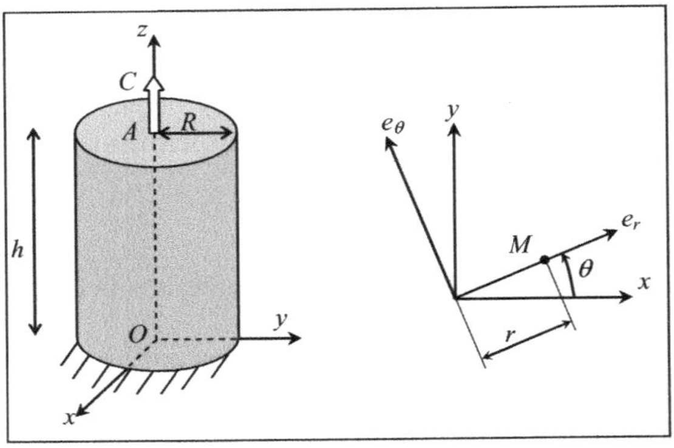

Figure 10.15. *Torsional cylinder*

QUESTION 2.–

We can determine the strain tensor by:

$$\underline{\underline{\varepsilon}} = \frac{1}{2}\cdot\left(\underline{\underline{grad}}(\underline{u}) + \underline{\underline{grad}}^t(\underline{u})\right) = \begin{bmatrix} 0 & 0 & 0 \\ 0 & 0 & \dfrac{k.r}{2} \\ 0 & \dfrac{k.r}{2} & 0 \end{bmatrix}_{(\underline{r},\underline{\theta},\underline{z})} \quad [10.82]$$

Be careful to use the gradient tensor in cylindrical coordinates for this relationship.

QUESTION 3.–

We can begin by observing that 0 is a principal strain and that $\underline{r}$ is the associated principal direction:

$$\varepsilon_I = 0 \quad \text{and} \quad \underline{x}_I = \underline{r} \quad [10.83]$$

Now we just have to work on the plane (θ, z) before determining the two principal strains and the associated principal directions. You will then find:

$$\begin{cases} \varepsilon_{II} = \dfrac{k.r}{2} \\ \varepsilon_{III} = -\dfrac{k.r}{2} \end{cases} \quad \text{and} \quad \begin{cases} \underline{x}_{II} = \underline{z} + \underline{\theta} \\ \underline{x}_{III} = \underline{z} - \underline{\theta} \end{cases} \quad [10.84]$$

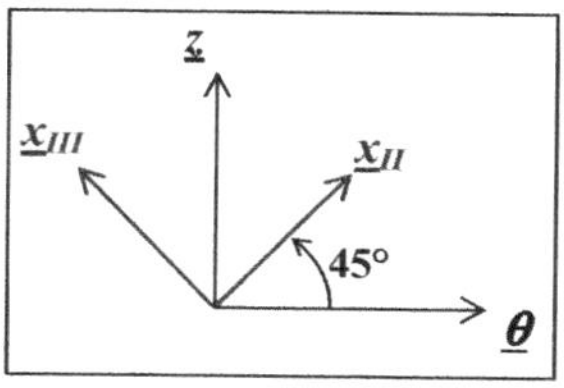

Figure 10.16. *Principal coordinate system. For a color version of this figure, see www.iste.co.uk/bouvet/aeronautical.zip*

QUESTION 4.–

Volume expansion (more generally known as volume variation) is equal to the trace of the strain tensor:

$$\frac{\Delta V}{V} = tr\left(\underline{\underline{\varepsilon}}\right) = 0 \qquad [10.85]$$

Since the trace is an invariant of the tensor, you can show that the result is the same irrespective of the coordinate system that you use for the calculation. In particular, here you can perform the calculation in (r, θ, z) or in (x_I, x_{II}, x_{III}).

Here we find that there is no volume variation. Since the displacement consists of the rotation of disks piled on top of each other, it is logical that the volume variation is zero.

In reality, we do see that the volume variation of a torsional cylinder is zero, but only at the first order. Indeed, we can show that there is a shortening of the cylinder but only at the second order. Since we perform all of the calculations here at the first order (in particular the relationship between the strain tensor and the displacement field gradient), we find no shortening and no volume variation.

QUESTION 5.–

The stress tensor is:

$$\underline{\underline{\sigma}} = 2.\mu.\underline{\underline{\varepsilon}} + \lambda.trace\left(\underline{\underline{\varepsilon}}\right).\underline{\underline{I}} = \begin{bmatrix} 0 & 0 & 0 \\ 0 & 0 & k.\mu.r \\ 0 & k.\mu.r & 0 \end{bmatrix}_{(\underline{r},\underline{\theta},\underline{z})} \qquad [10.86]$$

We then simply write out that the moment of the stress vector on the top side is equal to the applied torque:

$$\iint_{Sh} \underline{AM} \wedge \underline{\sigma}(M,\underline{z}).dS = C.\underline{z} \qquad [10.87]$$

where A is the center of the top side. With:

$$\iint_{Sh} \underline{AM} \wedge \underline{\sigma}(M,\underline{z}).dS = \iint_{Sh} \left(r.\underline{e}_r\right) \wedge \left(k.\mu.r.\underline{\theta}\right).r.d\theta.dr = k.\mu.\frac{R^4}{4}.2\pi.\underline{z} \qquad [10.88]$$

Hence:

$$k = \frac{2.C}{\pi.\mu.R^4} \quad [10.89]$$

And here we find the conventional relationship of strength of materials:

$$\tau_{\theta z} = \frac{C}{I_0}.r \quad [10.90]$$

where I_0 is the quadratic moment of inertia of a circular torsional beam:

$$I_0 = \iint_S r^2.dS = \frac{\pi.D^4}{32} \quad [10.91]$$

QUESTION 6.–

For the stress, strain and displacement fields to be the solution here, they must verify:

– The equilibrium equation:

You can verify that the divergence of the stress tensor is zero (not forgetting to take the divergence expression in cylindrical coordinates, noting that the volumic force is zero here).

– The relationship between stress and strain:

This relationship is clearly verified; moreover, this is how we determined the strain tensor according to the displacement field.

– The behavior law linking stress and strain:

This relationship is clearly verified; moreover, this is how we determined the stress tensor from the strain tensor.

– The stress and displacement boundary conditions:

The displacement boundary condition is the clamping of the lower side:

$$\text{In } z = 0,\ \underline{u}\big(M(r,\theta,0)\big) = \underline{0} \quad [10.92]$$

This relationship is verified automatically on seeing the shape of the displacement field selected.

The stress boundary conditions are divided in two, first for the lateral side, then for the top side:

For the lateral side, there is no external force, so we have:

$$\text{In } r = R\text{: } \underline{\underline{\sigma}}\left(M(R,\theta,z),\underline{n}_{ext}\right) = \underline{\underline{\sigma}}\left(M(R,\theta,z)\right).\underline{e}_r = \underline{0} \qquad [10.93]$$

This relationship is verified.

On the top side, the sum of the applied forces is a torque C in the z-direction. Nonetheless, we must not forget to prove that the force is indeed zero:

$$\begin{cases} \iint_{Sh} \underline{\underline{\sigma}}(M,\underline{z}).dS = \underline{0} \\ \iint_{Sh} \underline{AM} \wedge \underline{\underline{\sigma}}(M,\underline{z}).dS = C.\underline{z} \end{cases} \qquad [10.94]$$

The second relationship is true since it has been used previously to determine the k coefficient. As for the first relationship, you can easily show that it is true by calculation or simply sketch the stress vectors on the top side for visual evidence that the result is clear (upon adding all the arrows, we clearly find 0, on the other hand if we add their moments at point A we find torque C).

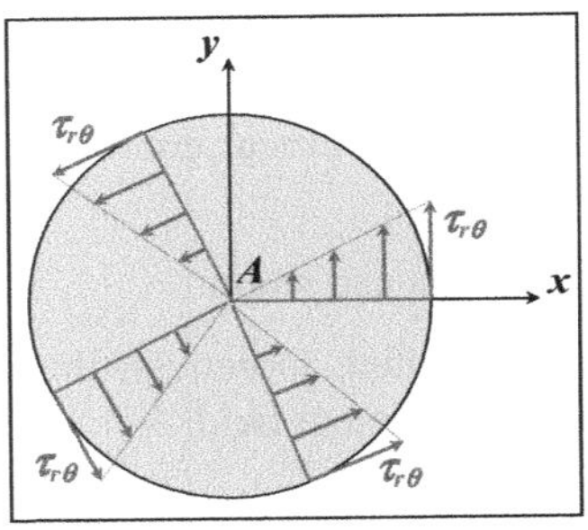

Figure 10.17. *Torsion stress. For a color version of this figure, see www.iste.co.uk/bouvet/aeronautical.zip*

QUESTION 7.–

We should begin by choosing a failure criterion adapted to a brittle material. The conventional criterion is that of maximum/minimum normal stress:

$$\begin{cases} Max\left(\sigma_I,\sigma_{II},\sigma_{III}\right) < \sigma_{trac} \\ Min\left(\sigma_I,\sigma_{II},\sigma_{III}\right) > \sigma_{comp} \end{cases} \qquad [10.95]$$

So we should start by determining the principal stresses, then find where they are maximum/minimum (and in what facet direction). Here, we can easily show that the principal stresses and directions are:

$$\begin{cases} \sigma_I = 0 \\ \sigma_{II} = k.\mu.r \\ \sigma_{III} = -k.\mu.r \end{cases} \quad \text{and} \quad \begin{cases} \underline{x}_I = \underline{r} \\ \underline{x}_{II} = \underline{z} + \underline{\theta} \\ \underline{x}_{III} = \underline{z} - \underline{\theta} \end{cases} \qquad [10.96]$$

We clearly see that the principal stress directions are the same as those for strain. This remains the case so long as the material is orthotropic (i.e. if the material has three perpendicular symmetry planes); which is particularly the case of an isotropic material for which all the planes are symmetrical!

Accordingly, the principal stresses are maximum/minimum where r is maximum, so here in $r = R$. Since a brittle material typically corresponds to greater resistance in compression than in tension (which is particularly the case of chalk, concrete, composite, glass, etc.), the criterion will be reached in $r = R$ and for a facet direction $\underline{x}_{II}$ and the crack will be perpendicular to $\underline{x}_{II}$. This crack will be shaped like a screw with a 45° angle to $\underline{z}$ (test this with a piece of chalk; a quick test beats a long explanation). Failure will be reached when the criterion is reached, that is:

$$Max(\sigma_I, \sigma_{II}, \sigma_{III}) = k.\mu.R = \frac{2.C}{\pi.R^3} = \sigma_{trac} \qquad [10.97]$$

Hence:

$$C = \frac{\pi.R^3.\sigma_{trac}}{2} \qquad [10.98]$$

QUESTION 8.–

We need to use the material to the maximum of its potential throughout the structure in order to decrease cylinder mass. In this example, we see that the material at the center of the cylinder is almost completely unnecessary. We should instead use a hollow cylinder with the greatest possible diameter (typically, this diameter is limited by the size of the structure and by the technical characteristics of its production). Where R_i and R_e are the internal and external diameters, we can show that:

$$Max(\sigma_I, \sigma_{II}, \sigma_{III}) = \frac{2.C.R_e}{\pi.(R_e^4 - R_i^4)} = \sigma_{trac} \qquad [10.99]$$

And where $e = R_e - R_i$ is the thickness of a hollow cylinder, assuming it to be small compared to R_e (allowing for a first-order limited development), we find:

$$Max(\sigma_I, \sigma_{II}, \sigma_{III}) = \frac{C}{2.\pi.R^2.e} = \sigma_{trac} \quad [10.100]$$

And since cylinder mass is proportional to $R.e$ (at the first order):

$$Mass = \rho.2.\pi.R.e.L \quad [10.101]$$

where ρ is the density and L is the cylinder length. Accordingly, we should minimize the $R.e$ ratio while keeping the $R^2.e$ ratio constant (as we may assume that C and σ_{trac} are provided). The radius must then be maximum in order to minimize mass:

$$R^2.e = cte \quad \Rightarrow \quad Mass = \rho.2.\pi.R.e.L = \frac{\rho.2.\pi.cte.L}{R} \quad [10.102]$$

10.8. Plastic compression

QUESTION 1.–

We see that stress tends to minus infinity when cube thickness tends to 0, but only if we write the behavior law with engineering stress/strain! Here, the law is written in true stress/strain. Between the true and engineering stress/strain, we have the relationship (see Chapter 7):

$$\begin{cases} \varepsilon_t = \ln(1+\varepsilon) \\ \sigma_t = \sigma.(1+\varepsilon) \end{cases} \quad [10.103]$$

where index t corresponds "true" and no index to "engineering". You will also see that I have omitted index t from the text of this exercise, as often happens in practice (and to push you to think!).

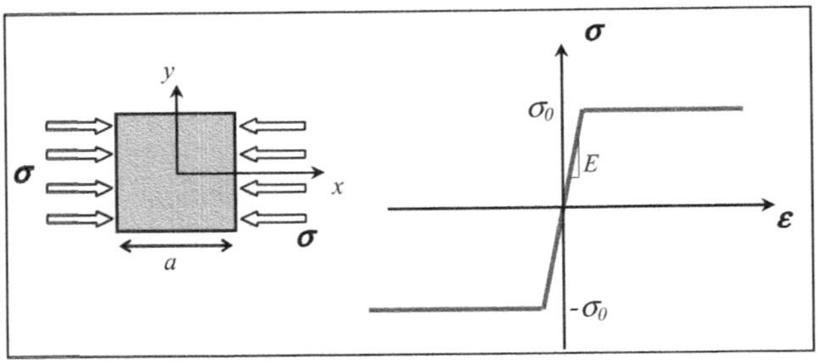

Figure 10.18. *Plastic compression. For a color version of this figure, see www.iste.co.uk/bouvet/aeronautical.zip*

If we plot this engineering stress/strain curve again, we obtain:

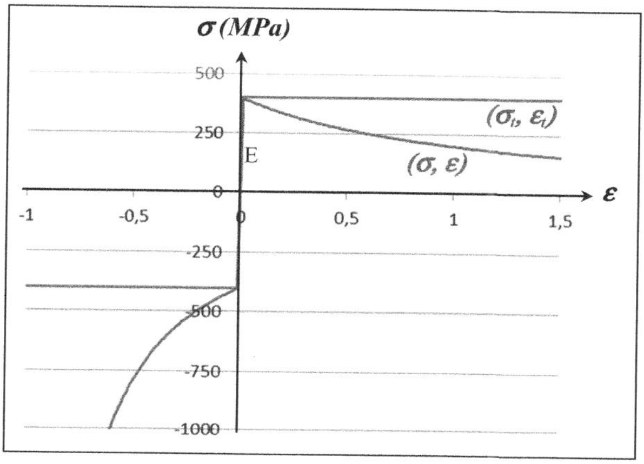

Figure 10.19. *True and engineering stress/strain tension curves. For a color version of this figure, see www.iste.co.uk/bouvet/aeronautical.zip*

As can be seen here, the elastic part of the curve is almost invisible when plotted to scale (for a 70 GPa modulus and an elastic limit of 400 MPa, the elastic strain limit is 0.00057).

Under compression, we see that if compression is continued indefinitely, then we obtain:

$$\begin{cases} \varepsilon_t \to -\infty \\ \sigma_t \to -\sigma_0 \end{cases} \quad \text{and} \quad \begin{cases} \varepsilon \to -1 \\ \sigma \to -\infty \end{cases} \qquad [10.104]$$

While under tension, if we continue tension indefinitely, then we obtain:

$$\begin{cases} \varepsilon_t \to +\infty \\ \sigma_t \to +\sigma_0 \end{cases} \quad \text{and} \quad \begin{cases} \varepsilon \to +\infty \\ \sigma \to 0 \end{cases} \qquad [10.105]$$

The force/displacement curve that you obtain will consequently be the same as the engineering stress/strain curve (except the cross-section for force and to the length for displacement).

QUESTION 2.–

In reality, when we perform a compression test, we obtain a barrel effect:

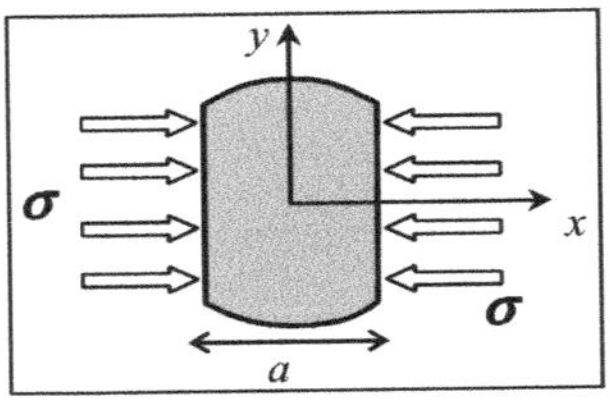

Figure 10.20. *Compressive barrel effect*

This effect is due to friction between the sample and the machine grips preventing the material from straining in the directions perpendicular to force (due to the Poisson effect). In practice, such barrel effect is very difficult to avoid, so we must do as much as possible to reduce the friction coefficient (even adding rollers to allow the material to strain freely, although this is an awkward procedure and can significantly disturb the stress field).

The solution of using a longer sample, as we do for tension, is clearly the wrong solution, since it leads to sample buckling!

QUESTION 3.–

If we perform a bi-compression test:

$$\underline{\underline{\sigma}} = \begin{bmatrix} \sigma & 0 & 0 \\ 0 & \sigma & 0 \\ 0 & 0 & 0 \end{bmatrix}_{(\underline{x},\underline{y},\underline{z})} \qquad \text{with } \sigma < 0 \qquad [10.106]$$

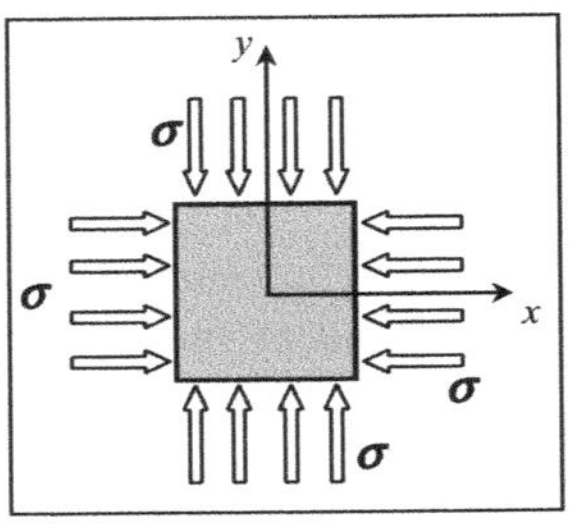

Figure 10.21. *Bi-compression*

We obtain the start of plasticity at the same stress as in pure tension. Indeed, if the material is ductile, then we can assume that its elasticity limit is correctly represented by Von Mises criterion:

$$\sigma_{VM} = \sqrt{\frac{1}{2}\cdot\left((\sigma_I - \sigma_{II})^2 + (\sigma_{II} - \sigma_{III})^2 + (\sigma_{III} - \sigma_I)^2\right)} = |\sigma| < \sigma_e \qquad [10.107]$$

And we find that the Von Mises stress is equal to the absolute value of σ, as with tension/compression.

QUESTION 4.–

One solution for performing this test consists of using beams to press on the cube. This experimental device allows us to limit barrel effect by leaving the material to strain freely during compression. In practice, this amounts to designing an orthotropic material with high longitudinal rigidity (relative to the beams' compression rigidity) compared to transversal rigidity, which is much lower (relative to beams' bending rigidity).

QUESTION 5.–

When performing a tri-compression test, plasticity is never reached (in theory at least!):

$$\underline{\underline{\sigma}} = \begin{bmatrix} \sigma & 0 & 0 \\ 0 & \sigma & 0 \\ 0 & 0 & \sigma \end{bmatrix}_{(\underline{x},\underline{y},\underline{z})} \quad \text{where } \sigma < 0 \qquad [10.108]$$

Indeed, we obtain zero Von Mises stress and never reach plasticity irrespective of the value of σ. This stems from the fact that plasticity is insensitive to hydrostatic pressure.

This does not mean that strain will be zero, since if plastic strain is zero, elastic strain will not be (and you can determine this strain).

10.9. Bi-material beam tension

Part 1: pure tension

QUESTION 1.–

With the beam under tension, it is logical to search for stress in both materials in a state of pure tension. Of course, this explanation is by no means a demonstration, but since we are aware that there is only one solution to this problem then if the proposed solution verifies all conditions (equilibrium equation, displacement and stress boundary conditions, compatibility condition, stress/strain relationship, displacement/stress relationship), it is certainly the correct solution.

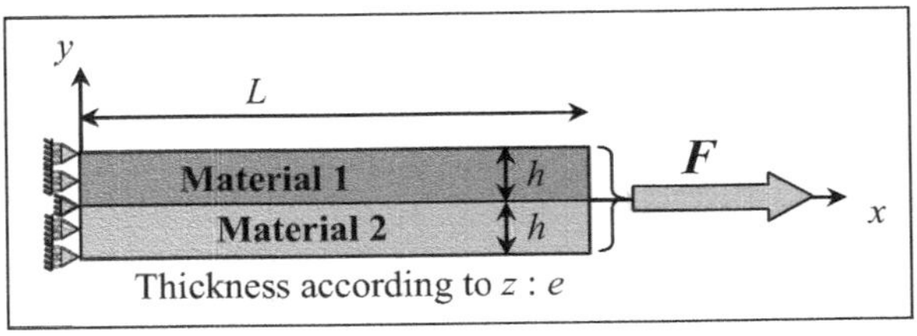

Figure 10.22. *Tension of a bi-material beam*

QUESTION 2.–

We write that the sum of the stress vectors on the side with the equation $x = L$ is equal to F:

$$- \text{in } x = L: \quad \iint_S \underline{\sigma}(M, \underline{x}).dS = \iint_S \underline{\underline{\sigma}}.\underline{x}.dS = F.\underline{x} \qquad [10.109]$$

For this calculation, we have to break down the integral on material 1 and material 2:

$$\iint_S \underline{\underline{\sigma}}.\underline{x}.dS = \iint_{S1} \underline{\underline{\sigma}}.\underline{x}.dS + \iint_{S2} \underline{\underline{\sigma}}.\underline{x}.dS = h.e.(a_1 + a_2).\underline{x} = F.\underline{x} \qquad [10.110]$$

Hence:

$$a_1 + a_2 = \frac{F}{h.e} \qquad [10.111]$$

QUESTION 3.–

Assuming that the material is linear elastic, homogeneous and isotropic, we find:

$$\underline{\underline{\varepsilon}}_1 = \begin{bmatrix} \frac{a_1}{E_1} & 0 \\ 0 & \frac{-\nu_1.a_1}{E_1} \end{bmatrix}_{(\underline{x},\underline{y})} \quad \text{and} \quad \underline{\underline{\varepsilon}}_2 = \begin{bmatrix} \frac{a_2}{E_2} & 0 \\ 0 & \frac{-\nu_2.a_2}{E_2} \end{bmatrix}_{(\underline{x},\underline{y})} \qquad [10.112]$$

QUESTION 4.–

If the materials are bonded at $y = 0$, then elongations in the x-direction, that is ε_x, must be equal (we will return to this condition later), so:

$$\varepsilon_{x1} = \varepsilon_{x2} \Rightarrow \frac{a_1}{E_1} = \frac{a_2}{E_2} \quad [10.113]$$

Hence:

$$\begin{cases} a_1 = \dfrac{F.E_1}{h.e.(E_1 + E_2)} \\ a_2 = \dfrac{F.E_2}{h.e.(E_1 + E_2)} \end{cases} \quad [10.114]$$

Accordingly, strain must be continuous at the bonding point, whereas stress is discontinuous:

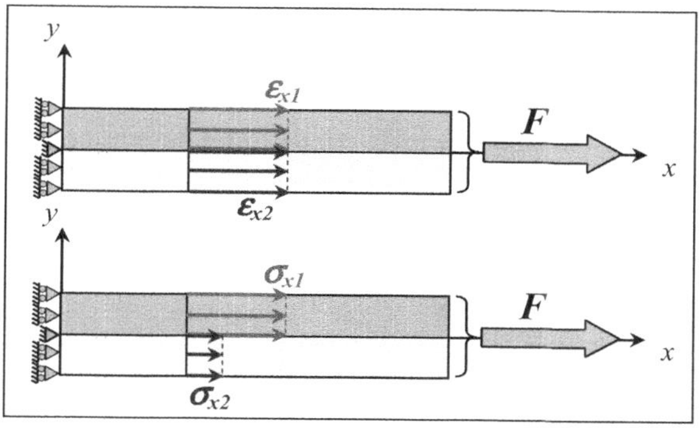

Figure 10.23. *Stress and strain through bi-material tension. For a color version of this figure, see www.iste.co.uk/bouvet/aeronautical.zip*

However, in reality this stress discontinuity does not occur. Stress can go from stress 1 to stress 2 over a very short distance, over the bonding interface between the two materials for instance. This bonding interface is particularly solicited, indeed, we can show that in large part this is due to τ_{xy} shear stress. Moreover, if this τ_{xy} shear stress is too great it will lead to the glue failure.

QUESTION 5.–

Since the stress and strain fields in materials 1 and 2 are different, we must find different displacement fields for both materials; $\underline{u}_1$ for material 1 and $\underline{u}_2$ for material 2. We will start with $\underline{u}_1(u_1, v_1)$:

$$\begin{cases} \varepsilon_{x1} = \dfrac{a_1}{E_1} = \dfrac{\partial u_1}{\partial x} \\ \varepsilon_{y1} = \dfrac{-\nu_1.a_1}{E_1} = \dfrac{\partial v_1}{\partial y} \\ \gamma_{xy1} = 0 = \dfrac{\partial u_1}{\partial y} + \dfrac{\partial v_1}{\partial x} \end{cases} \quad [10.115]$$

By integrating the first two equations, we find:

$$\begin{cases} u_1 = \dfrac{a_1}{E_1}.x + f(y) \\ v_1 = \dfrac{-\nu_1.a_1}{E_1}.y + g(x) \end{cases} \quad [10.116]$$

where $f(y)$ and $g(x)$ are two unknown functions to be determined. Reinsert this into the third equation, and we find:

$$\frac{df}{dy}(y) + \frac{dg}{dx}(x) = 0 \quad \Rightarrow \quad \frac{df}{dy}(y) = -\frac{dg}{dx}(x) = ct_1 \quad [10.117]$$

Indeed, if two functions dependent on two different variables are equal, then they must be constant.

Hence:

$$\begin{cases} u_1 = \dfrac{a_1}{E_1}.x + ct_1.y + ct_2 \\ v_1 = \dfrac{-\nu_1.a_1}{E_1}.y - ct_1.x + ct_3 \end{cases} \quad [10.118]$$

In order to determine these three constants, we must use the boundary conditions. In practice, the $\underline{u}_0$ vector (ct_2, ct_3) represents a translation of a rigid-body displacement field and ct_1 a rotation around $\underline{z}$. You can show this result by determining the displacement field induced by this rotation:

$$\underline{\Omega} = -ct_1.\underline{z} \qquad [10.119]$$

By:

$$\underline{u}_{\Omega} = \underline{\Omega} \wedge \underline{OM} = \begin{bmatrix} 0 \\ 0 \\ -cte_1 \end{bmatrix} \wedge \begin{bmatrix} x \\ y \\ 0 \end{bmatrix} = \begin{bmatrix} ct_1.y \\ -ct_1.x \\ 0 \end{bmatrix} \qquad [10.120]$$

Essentially, any of the displacement fields equal to a rigid-body displacement field can be a solution here. In other words, if a displacement field is a solution, then adding the rigid-body displacement field changes nothing, since its derivative, and then the strain field, remains the same.

In this instance, since:

$$\begin{cases} u_1(0,y) = 0 \\ v_1(0,0) = 0 \end{cases} \qquad [10.121]$$

Then:

$$\begin{cases} u_1 = \dfrac{a_1}{E_1}.x \\ v_1 = \dfrac{-\nu_1.a_1}{E_1}.y \end{cases} \qquad [10.122]$$

And going through the same process for material 2, we find:

$$\begin{cases} u_2(x,y) = \dfrac{a_2}{E_2}.x = \dfrac{a_1}{E_1}.x = u_1(x,y) \\ v_2(x,y) = \dfrac{-\nu_2.a_2}{E_2}.y = \dfrac{-\nu_1.a_1}{E_1}.y = v_1(x,y) \end{cases} \qquad [10.123]$$

And here we see that the two displacement fields are equal. In the general case, we ought to show that the displacement fields are equal at the bonding interface (as is the case here) but not necessarily everywhere.

QUESTION 6.–

We must start by defining what the mean modulus is. Suppose I had given you the bi-material without saying that it is made of two different materials, and asked you to determine its Young's modulus. You would then have performed a tension test, determining stress by $\overline{\sigma_x} = F / S$ and strain by $\overline{\varepsilon_x} = \Delta L / L$ (where the $\overline{x}$ notation indicates mean of x), then plotting the stress curve according to strain. The slope of this curve would have been the mean modulus. So, here:

$$\begin{cases} \overline{\sigma_x} = \dfrac{F}{S} = \dfrac{F}{2.h.e} = \dfrac{a_1 + a_2}{2} = \dfrac{\sigma_{x1} + \sigma_{x2}}{2} \\ \overline{\varepsilon_x} = \dfrac{\Delta L}{L} = \dfrac{u_1(x = L)}{L} = \dfrac{a_1}{E_1} = \dfrac{a_2}{E_2} = \dfrac{\varepsilon_{x1} + \varepsilon_{x2}}{2} = \varepsilon_{x1} = \varepsilon_{x2} \end{cases} \quad [10.124]$$

In passing, we find that the mean stress is the mean of the two stresses and that the strains are all the same.

Hence the mean Young's modulus, which here is clearly equal to the mean of Young's moduli:

$$\overline{E} = \frac{\overline{\sigma_x}}{\overline{\varepsilon_x}} = \frac{E_1 + E_2}{2} \quad [10.125]$$

QUESTION 7.–

On review, we see that the stress, strain and displacement fields are almost the solution to this problem:

– the equilibrium equation is verified (I will leave you to do the calculation, but since the stress field is constant);

– the stress/strain relationship is verified;

– the strain/displacement relationship is verified;

– the displacement boundary conditions are verified (at $x = 0$);

– the bonding interface conditions at $y = 0$ are verified.

To be more precise, we should verify that the displacement fields of both materials are equal at the bonding interface, as is the case here. You should also note that this condition leads to strains in the x-direction being equal as well (which is only true since the bond is in the x-direction).

We should also verify that the stress vectors for the bonding plane facets are equal for both materials (otherwise the bonding interface would not be in equilibrium!). This is indicated here by $\underline{\sigma}_1\left(M(x,0),-\underline{y}\right)=\underline{\sigma}_2\left(M(x,0),\underline{y}\right)$, which is verified.

The stress boundary conditions are almost verified. Indeed, you can show that in $y = \pm h$, the stress vector is equal to external force (which is zero here).

We must then show that the sum of the stress vectors on the force application side (in $x = L$) is equal to F. As a result, this relationship is verified (it is the relationship that we used to determine stress):

$$-\text{ in } x = L: \quad \iint_S \underline{\sigma}(M,\underline{x}).dS = \iint_S \underline{\underline{\sigma}}.\underline{x}.dS = F.\underline{x} \qquad [10.126]$$

However, we have not verified the moment equation. Here the problem becomes more challenging. Clearly, if we apply force at the middle of the side (at $x = L$ and $y = 0$), it will not work:

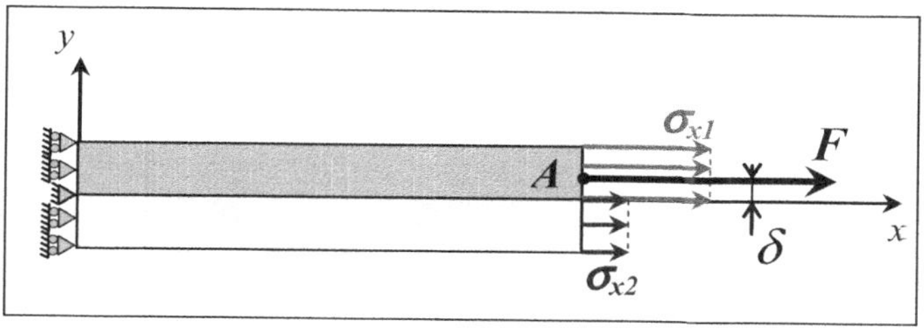

Figure 10.24. *Stress at the end of a bi-material beam under tension. For a color version of this figure, see www.iste.co.uk/bouvet/aeronautical.zip*

So, the force F must necessarily be moved by a distance δ for there to be in equilibrium. We can determine this distance by writing that the moment is zero at this point:

$$-\text{ in } x = L: \quad \iint_S \underline{AM} \wedge \underline{\sigma}(M,\underline{x}).dS = \underline{0} \qquad [10.127]$$

Then:

$$\iint_S \underline{AM} \wedge \underline{\sigma}(M,\underline{x}).dS = \iint_{S1} \underline{AM} \wedge \underline{\sigma}_1(M,\underline{x}).dS$$
$$+ \iint_{S2} \underline{AM} \wedge \underline{\sigma}_2(M,\underline{x}).dS = \underline{0}$$
$$= \int_0^h (y-\delta).\sigma_{x1}.e.dy + \int_{-h}^0 (y-\delta).\sigma_{x2}.e.dy$$

$$= \left(\frac{h^2}{2} - \delta.h\right).E_1 + \left(-\frac{h^2}{2} - \delta.h\right).E_2 \qquad [10.128]$$

Hence:

$$\delta = \frac{(E_1 - E_2).h}{2.(E_1 + E_2)} \qquad [10.129]$$

If, in addition, we take $E_1 = 3.E_2$, then we find $\delta = h/4$.

In conclusion, if you can apply force at point A, then this will indeed be the solution here. In practice, this is done by restraining the rotation around $\underline{z}$ at the machine grip, and then generating a moment (which is equivalent to displacing the force application point):

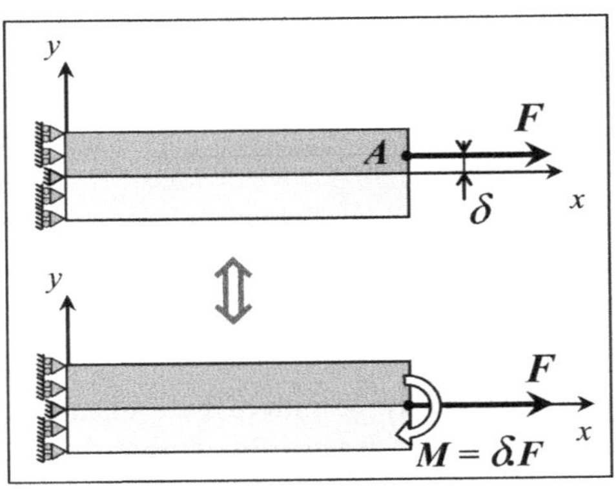

Figure 10.25. *Bi-material beam tension test. For a color version of this figure, see www.iste.co.uk/bouvet/aeronautical.zip*

If we now apply F right in the middle (at $y = 0$), the solution above is no longer the solution to this problem, and you have to continue the exercise!

Part 2: tension/bending

QUESTION 8.–

Now we need to apply F at exactly the point where $y = 0$.

Here, the different rigidity of the two materials will induce bending moment in the beam in the z-direction:

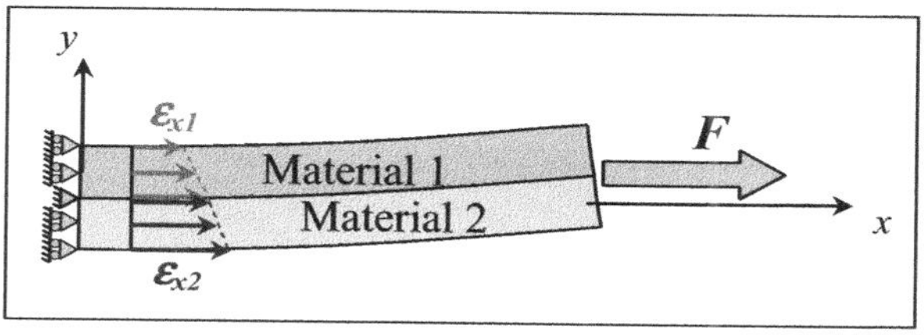

Figure 10.26. *Strain during bi-material tension/bending. For a color version of this figure, see www.iste.co.uk/bouvet/aeronautical.zip*

In physical terms, this is quite easy to understand: if material 1 is more rigid than material 2, i.e. more difficult to strain, then it will be less strained than material 2 (nature obeys the law of least effort, as do students, though perhaps not if you have come this far!).

So, we can assume that the strain field will be a bending strain field:

$$\varepsilon_{x1} = \varepsilon_{x2} = a + b.y \qquad [10.130]$$

So, we have:

$$\begin{cases} \sigma_{x1} = E_1.(a+b.y) = 3.E_2.(a+b.y) \\ \sigma_{x2} = E_2.(a+b.y) \end{cases} \qquad [10.131]$$

You can then write that force at $x = L$ is F:

$$\text{in } x = L: \iint_S \underline{\underline{\sigma}}(M,\underline{x}).dS = \iint_S \underline{\underline{\sigma}}.\underline{x}.dS = \int_0^h \sigma_{x1}.e.dy + \int_{-h}^0 \sigma_{x2}.e.dy = F.\underline{x} \qquad [10.132]$$

hence:

$$E_2.(4.a + b.h) = \frac{F}{h.e} \qquad [10.133]$$

And that moment at $x = L$ and $y = 0$ is zero:

$$\begin{aligned} \iint_S \underline{AM} \wedge \underline{\sigma}(M,\underline{x}).dS &= \iint_{S1} \underline{AM} \wedge \underline{\sigma}_1(M,\underline{x}).dS \\ &\quad + \iint_{S2} \underline{AM} \wedge \underline{\sigma}_2(M,\underline{x}).dS = \underline{0} \\ &= \int_0^h y.\sigma_{x1}.e.dy + \int_{-h}^0 y.\sigma_{x2}.e.dy \end{aligned} \qquad [10.134]$$

hence:

$$a = \frac{-4}{3}.b.h \qquad [10.135]$$

And finally:

$$\begin{cases} a = \dfrac{4.F}{13.e.E_2.h} \\ b = \dfrac{-3.F}{13.e.E_2.h} \end{cases} \qquad [10.136]$$

which gives a rather complex stress field:

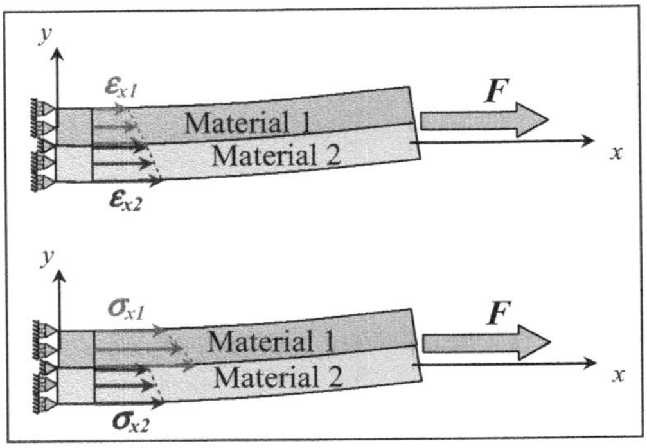

Figure 10.27. *Stress and strain during bi-material tension/bending. For a color version of this figure, see www.iste.co.uk/bouvet/aeronautical.zip*

So, strain is continuous through the thickness (and with linear variation), while strain is discontinuous at the bonding interface (and with linear variation).

I will leave you to show that all of these conditions are verified and that this is indeed the solution here.

10.10. Beam thermal expansion

Part 1: Mono-material beam

QUESTION 1.–

If the beam is assumed to be constrained at both ends, then the strain field in the x-direction will be zero (there will be no displacement in the x-direction) and we can assume that the strain field will be in the form of a pure tension strain field (with $\gamma_{xy} = 0$ in particular):

$$\underline{\underline{\varepsilon}} = \begin{bmatrix} 0 & 0 \\ 0 & \varepsilon_y \end{bmatrix}_{(\underline{x},\underline{y})} \quad [10.137]$$

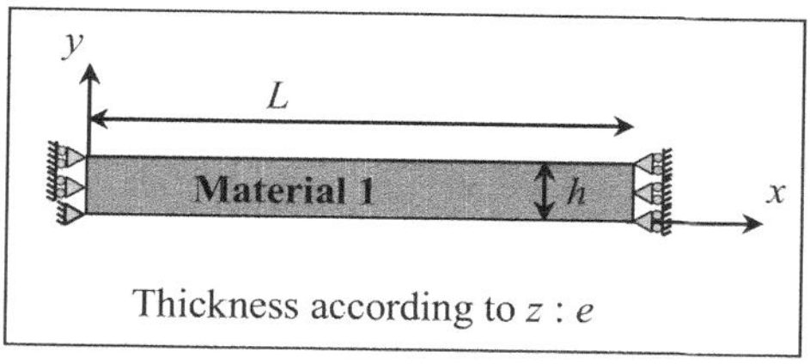

Figure 10.28. *Thermal expansion of a constrained beam*

However, at the end of the exercise we must remember to show that all of the conditions were met and that we do indeed have the solution here.

The stress field will also be one of pure tension:

$$\underline{\underline{\sigma}} = \begin{bmatrix} \sigma_x & 0 \\ 0 & 0 \end{bmatrix}_{(\underline{x},\underline{y})} \quad [10.138]$$

The thermal strain field will be that due to thermal expansion:

$$\underline{\underline{\varepsilon_{th}}} = \begin{bmatrix} \alpha.\Delta T & 0 \\ 0 & \alpha.\Delta T \end{bmatrix}_{(\underline{x},\underline{y})} \quad [10.139]$$

Finally, the behavior relationship leads to:

$$\underline{\underline{\varepsilon_e}} = \underline{\underline{\varepsilon}} - \underline{\underline{\varepsilon_{th}}} = \frac{1+\nu}{E}\underline{\underline{\sigma}} - \frac{\nu}{E}.trace\left(\underline{\underline{\sigma}}\right).\underline{\underline{I}} \Rightarrow \begin{cases} 0 = \dfrac{\sigma_x}{E} + \alpha.\Delta T \\ \varepsilon_y = \dfrac{-\nu.\sigma_x}{E} + \alpha.\Delta T \end{cases} \quad [10.140]$$

Hence:

$$\begin{cases} \sigma_x = -\alpha.E.\Delta T = -252\,MPa \\ \varepsilon_y = \alpha.(1+\nu).\Delta T = 1560\,\mu\varepsilon \end{cases} \quad [10.141]$$

The elongation in the y-direction is positive due to thermal expansion, and stress in the x-direction is a compressive stress due to impeded expansion (essentially, the material would have expanded, but since it has been prevented from doing so, compression appears instead).

You can verify that the stress and strain fields proposed are indeed the solution here.

QUESTION 2.–

If we assume the beam is free, then the stress field will be zero:

$$\underline{\underline{\sigma}} = \begin{bmatrix} 0 & 0 \\ 0 & 0 \end{bmatrix}_{(\underline{x},\underline{y})} \quad [10.142]$$

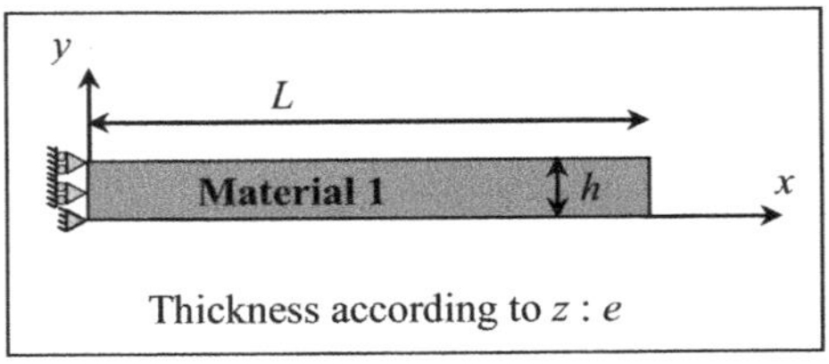

Figure 10.29. *Thermal expansion of a free beam*

The thermal strain field will be that due to thermal expansion:

$$\underline{\underline{\varepsilon}}_{th} = \begin{bmatrix} \alpha.\Delta T & 0 \\ 0 & \alpha.\Delta T \end{bmatrix}_{(\underline{x},\underline{y})} \qquad [10.143]$$

finally, the behavior relationship leads to:

$$\begin{aligned} \underline{\underline{\varepsilon}} = \underline{\underline{\varepsilon}}_e + \underline{\underline{\varepsilon}}_{th} &= \frac{1+\nu}{E}\underline{\underline{\sigma}} - \frac{\nu}{E}.trace\left(\underline{\underline{\sigma}}\right).\underline{\underline{I}} + \alpha.\Delta T.\underline{\underline{I}} \\ &= \begin{bmatrix} \alpha.\Delta T & 0 \\ 0 & \alpha.\Delta T \end{bmatrix}_{(\underline{x},\underline{y})} = \begin{bmatrix} 1200 & 0 \\ 0 & 1200 \end{bmatrix}_{(\underline{x},\underline{y})} (in\ \mu\varepsilon) \end{aligned} \qquad [10.144]$$

In conclusion, the material needs to dilate, so we leave it to dilate, and doing so it does not generate stress.

Again, you can verify that the stress and strain fields proposed are indeed the solution here.

Part 2: bi-material beam

QUESTION 3.–

For total strain, we have:

$$\underline{\underline{\varepsilon}}_1 = \begin{bmatrix} 0 & 0 \\ 0 & \varepsilon_{y1} \end{bmatrix}_{(\underline{x},\underline{y})} \quad \text{and} \quad \underline{\underline{\varepsilon}}_2 = \begin{bmatrix} 0 & 0 \\ 0 & \varepsilon_{y2} \end{bmatrix}_{(\underline{x},\underline{y})} \qquad [10.145]$$

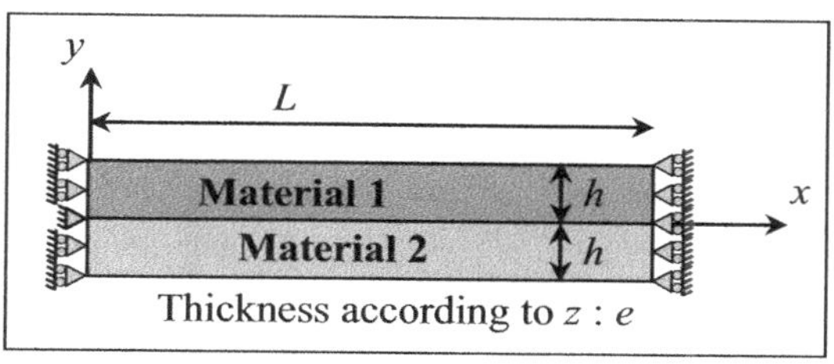

Figure 10.30. *Thermal expansion of a constrained bi-material beam*

And for stress:

$$\underline{\underline{\sigma}}_1 = \begin{bmatrix} \sigma_{x1} & 0 \\ 0 & 0 \end{bmatrix}_{(\underline{x},\underline{y})} \quad \text{and} \quad \underline{\underline{\sigma}}_2 = \begin{bmatrix} \sigma_{x2} & 0 \\ 0 & 0 \end{bmatrix}_{(\underline{x},\underline{y})} \qquad [10.146]$$

So, the thermal strain field is that due to thermal expansion:

$$\underline{\underline{\varepsilon_{th}}}_1 = \begin{bmatrix} \alpha_1.\Delta T & 0 \\ 0 & \alpha_1.\Delta T \end{bmatrix}_{(\underline{x},\underline{y})} \quad \text{and} \quad \underline{\underline{\varepsilon_{th}}}_2 = \begin{bmatrix} \alpha_2.\Delta T & 0 \\ 0 & \alpha_2.\Delta T \end{bmatrix}_{(\underline{x},\underline{y})} \qquad [10.147]$$

Having applied the behavior law in both materials, we have:

$$\begin{cases} \sigma_{x1} = -\alpha_1.E_1.\Delta T = -252\,MPa \\ \sigma_{x2} = -\alpha_2.E_2.\Delta T = -168\,MPa \\ \varepsilon_{y1} = \alpha_1.(1+\nu_1).\Delta T = 1560\,\mu\varepsilon \\ \varepsilon_{y2} = \alpha_2.(1+\nu_2).\Delta T = 3120\,\mu\varepsilon \end{cases} \qquad [10.148]$$

In conclusion, there is no difference to the case of a single material; the two materials act as they would without interaction! This is due to the fact that the bonding conditions (displacements and stress vectors have to be equal at the bonding interface) are verified with the stress and strain fields of a single material.

QUESTION 4.–

The problem is more complex in the case of the free beam.

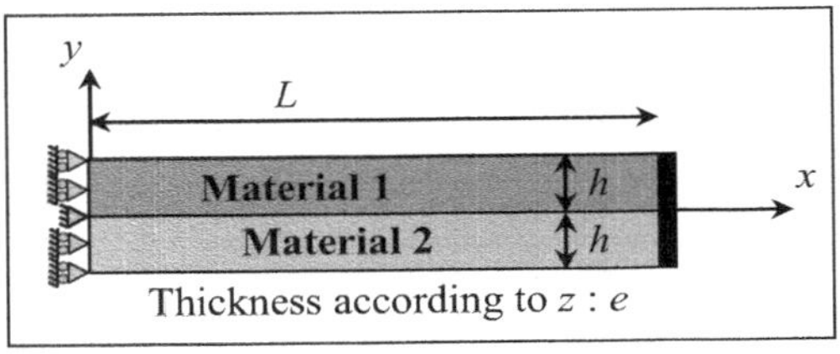

Figure 10.31. *Thermal expansion of a free bi-material beam*

In this case, if you follow the same logic as above, you should find:

$$\begin{cases} \sigma_{x1} = 0 \\ \sigma_{x2} = 0 \\ \varepsilon_{x1} = \varepsilon_{y1} = \alpha_1.\Delta T = 1200\,\mu\varepsilon \\ \varepsilon_{x2} = \varepsilon_{y2} = \alpha_2.\Delta T = 2400\,\mu\varepsilon \end{cases} \quad [10.149]$$

But the bonding condition is not verified here! Indeed, if the strains in the x-direction were different for the two materials, then the elongation of the bonding interface would be different for both, leading to the bond's failure!

In reality, the difference in thermal expansion coefficients for the two materials induces bending in the beam:

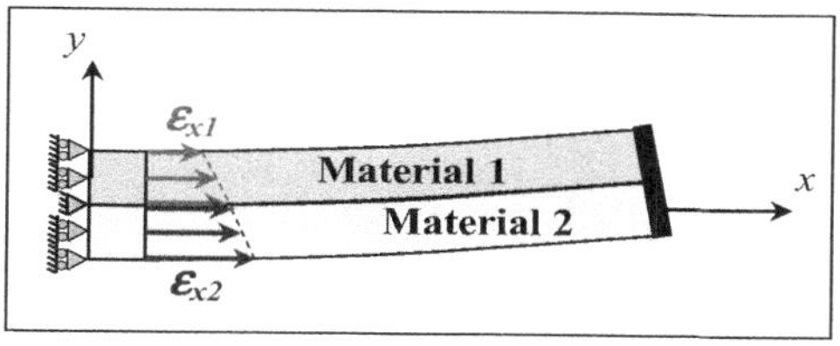

Figure 10.32. *Strain due to thermal expansion of a free bi-material beam. For a color version of this figure, see www.iste.co.uk/bouvet/aeronautical.zip*

So, we can take these strain fields in the form:

$$\underline{\underline{\varepsilon}}_1 = \begin{bmatrix} a+b.y & 0 \\ 0 & \varepsilon_{y1} \end{bmatrix}_{(\underline{x},\underline{y})} \quad \text{and} \quad \underline{\underline{\varepsilon}}_2 = \begin{bmatrix} a+b.y & 0 \\ 0 & \varepsilon_{y2} \end{bmatrix}_{(\underline{x},\underline{y})} \quad [10.150]$$

Once again, this is not a demonstration; you should ultimately show that with this strain field all of the problem's conditions are verified and that your solution is indeed the solution here.

And we still have:

$$\underline{\underline{\sigma}}_1 = \begin{bmatrix} \sigma_{x1} & 0 \\ 0 & 0 \end{bmatrix}_{(\underline{x},\underline{y})} \quad \text{and} \quad \underline{\underline{\sigma}}_2 = \begin{bmatrix} \sigma_{x2} & 0 \\ 0 & 0 \end{bmatrix}_{(\underline{x},\underline{y})} \quad [10.151]$$

where stresses in the x-direction are *a priori* functions of y.

So, having applied the behavior law in both materials:

$$\begin{cases} a+b.y=\dfrac{\sigma_{x1}}{E_1}+\alpha_1.\Delta T \\ a+b.y=\dfrac{\sigma_{x2}}{E_2}+\alpha_2.\Delta T \\ \varepsilon_{y1}=\dfrac{-\nu.\sigma_{x1}}{E_1}+\alpha_1.\Delta T \\ \varepsilon_{y2}=\dfrac{-\nu.\sigma_{x2}}{E_2}+\alpha_2.\Delta T \end{cases} \quad \text{(taking } \nu_1=\nu_2=\nu\text{)} \qquad [10.152]$$

Hence:

$$\begin{cases} \sigma_{x1}=3.E_2.(a-\alpha_1.\Delta T+b.y) \\ \sigma_{x2}=E_2.(a-\alpha_2.\Delta T+b.y) \end{cases} \quad \text{(taking } E_1=3.E_2\text{)} \qquad [10.153]$$

We must then write that the force and moment at the end of the beam is zero:

– in $x = L$ and at point $A\ (L,0)$:

$$\begin{cases} \iint_S \underline{\sigma}(M,\underline{x}).dS=\int_0^h \sigma_{x1}.e.dy+\int_{-h}^0 \sigma_{x2}.e.dy=\underline{0} \\ \iint_S \underline{AM}\wedge\underline{\sigma}(M,\underline{x}).dS=\int_0^h y.\sigma_{x1}.e.dy+\int_{-h}^0 y.\sigma_{x2}.e.dy=\underline{0} \end{cases} \qquad [10.154]$$

So you have two equations and two unknowns (a and b), hence:

$$\begin{cases} a=\dfrac{(15.\alpha_1+11.\alpha_2).\Delta T}{26}=1.71.10^{-3} \\ b=\dfrac{9.(\alpha_1-\alpha_2).\Delta T}{13.h}=-1.67.10^{-4}\ mm^{-1} \end{cases} \qquad [10.155]$$

Accordingly, strain is continuous through the thickness (and with linear variation) while stress is discontinuous at the bonding interface (and with linear variation):

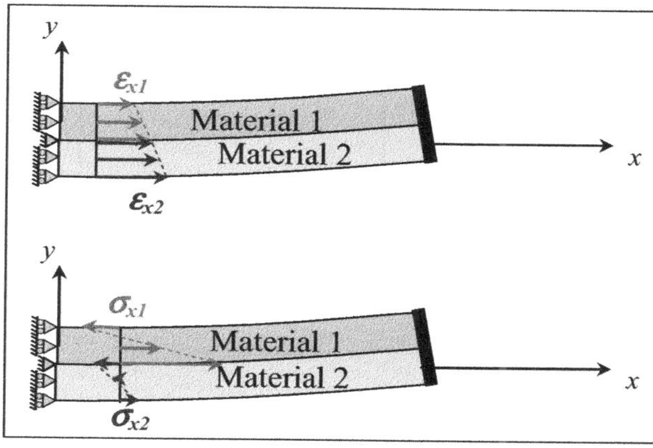

Figure 10.33. *Stress and strain field due to thermal expansion of a free bi-material beam. For a color version of this figure, see www.iste.co.uk/bouvet/aeronautical.zip*

In order to verify your calculations, you can show that:

$$\begin{cases} \varepsilon_{x1}(y=h)=857\ \mu\varepsilon \\ \varepsilon_{x2}(y=-h)=2550\ \mu\varepsilon \\ \sigma_{x1}(y=h)=-68\ MPa \\ \sigma_{x1}(y=0)=101\ MPa \\ \sigma_{x2}(y=0)=-48\ MPa \\ \sigma_{x2}(y=-h)=10\ MPa \end{cases} \quad [10.156]$$

Once again, you can verify that the stress and strain fields proposed are indeed the solution here.

10.11. Cube under shear stress

QUESTION 1.–

By writing that the sum of moments is zero (at point O for instance), you will find:

$$M = F.a \quad [10.157]$$

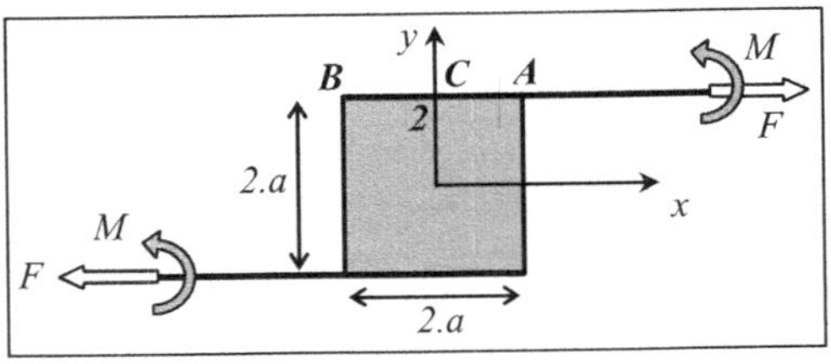

Figure 10.34. *Cube under shear stress*

QUESTION 2.–

We take:

$$\underline{\underline{\sigma}} = \begin{bmatrix} 0 & K.(a^2 - x^2) \\ K.(a^2 - x^2) & 2.K.x.y \end{bmatrix}_{(\underline{x},\underline{y})} \quad [10.158]$$

For this tensor to be the solution here, we must verify:

– the equilibrium equation. I will leave you to verify that the divergence of this tensor is indeed zero;

– stress boundary conditions:

- in $x = \pm a$: $\underline{\sigma}(M(\pm a, y), \underline{n}_{ext}) = \underline{\underline{\sigma}}(M(\pm a, y)).\pm\underline{x} = \underline{0}$, which is indeed verified;

- in $y = a$: $\begin{cases} \iint_S \underline{\sigma}(M(x,a), \underline{y}).dS = F.\underline{x} \\ \iint_S \underline{CM} \wedge \underline{\sigma}(M(x,a), \underline{y}).dS = M.\underline{z} \end{cases}$

where the point C is the center of the side with equation $y = a$. These conditions mean that the sum of all the stress vectors on this side must be equal to the applied force (in force and in moment).

Hence:

$$\begin{cases} \iint_S \tau_{xy}(x,a).dS = F \\ \iint_S \sigma_y(x,a).dS = 0 \\ \iint_S x.\sigma_y(x,a).dS = M \end{cases} \quad [10.159]$$

We can show that the second equation is verified automatically, and that the other two lead to the same result, thus allowing us to determine K according to force F:

$$K = \frac{3.F}{8.a^4} \qquad [10.160]$$

You can show that the stress boundary conditions in $y = -a$ lead to the same result (the problem is symmetrical!).

We must then determine the strain tensor, integrating it to determine the displacement field, then show that the displacement boundary conditions are verified. The problem here is not entirely explicit, and these conditions are not provided, which is not too significant since here we are mostly interested in the stress field.

And we must verify that the proposed stress field indeed verified the compatibility condition:

$$\underline{\underline{\Delta\sigma}} + \frac{1}{1+\nu}.\underline{\underline{grad}}\left(\underline{grad}\left(trace\left(\underline{\underline{\sigma}}\right)\right)\right) = \underline{\underline{0}} \qquad [10.161]$$

I leave you to verify that this is indeed the case here.

QUESTION 3.–

Plotting stress vectors on the external faces allow us to understand how the material is strained and to image the stress boundary conditions (here M induces σ_y and F induces τ_{xy}):

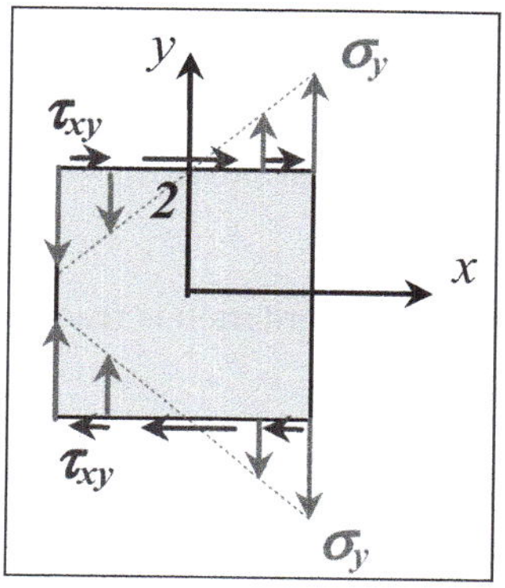

Figure 10.35. *Stress vectors on a cube under shear stress. For a color version of this figure, see www.iste.co.uk/bouvet/aeronautical.zip*

QUESTION 4.–

Since the material is brittle, we can use a maximum/minimum normal stress criterion (remember that the maximum/minimum normal stress is necessarily one of the three principal stresses):

$$\begin{cases} Max(\sigma_I,\sigma_{II},\sigma_{III}) < \sigma_{trac} \\ Min(\sigma_I,\sigma_{II},\sigma_{III}) > \sigma_{comp} \end{cases} \quad [10.162]$$

Since here the problem is 2D, we clearly have 0 as the third principal stress and $\underline{z}$ the associated principal direction.

We must then determine the other two principal stresses, so here:

$$\sigma_{I,II} = K.\left(x.y \pm \sqrt{(x.y)^2 + (a^2 - x^2)^2}\right) \quad [10.163]$$

You can show that these two expressions are maximum/minimum at the four corners of the square. We must simply study the stress state at these four points; and even at the two top points due to the problem's symmetry:

$$\begin{cases} \underline{\underline{\sigma}}(A(a,a)) = \begin{bmatrix} 0 & 0 \\ 0 & 2.K.a^2 \end{bmatrix}_{(\underline{x},\underline{y})} \\ \underline{\underline{\sigma}}(B(-a,a)) = \begin{bmatrix} 0 & 0 \\ 0 & -2.K.a^2 \end{bmatrix}_{(\underline{x},\underline{y})} \end{cases} \quad [10.164]$$

Since brittle materials typically have tensile stress less than that of compression (in absolute value), failure will occur at point A, with the crack developing in the perpendicular direction to $\underline{y}$.

Failure will occur when the failure criterion is reached, so:

$$2.K.a^2 = \sigma_{trac} \quad \Rightarrow \quad F = \frac{4}{3}.a^2.\sigma_{trac} \quad [10.165]$$

QUESTION 5.–

If the material is ductile, we can use the Tresca criterion (which is a criterion for the start of plasticity rather than a failure criterion!):

$$\sigma_{tresca} = 2.\tau_{\max} = Max\left(\left|\sigma_I - \sigma_{II}\right|;\left|\sigma_{II} - \sigma_{III}\right|;\left|\sigma_{III} - \sigma_I\right|\right) < \sigma_e \qquad [10.166]$$

Here, we obtain:

$$\sigma_{tresca} = 2.\tau_{\max} = 2.K.\sqrt{(x.y)^2 + (a^2 - x^2)^2} \qquad [10.167]$$

And we can show that this criterion is maximum at A, B and C (and at the three symmetrical points with y negative) for:

$$2.K.a^2 = \sigma_e \quad \Rightarrow \quad F = \frac{4}{3}.a^2.\sigma_e \qquad [10.168]$$

So plasticity will begin at the same time at the three points A, B and C (and at three symmetrical points with y negative). In the case of a ductile material, plasticity will then spread outwards from these six points but without leading to failure (at least not immediately).

10.12. Spherical reservoir under pressure

QUESTION 1.–

A priori the displacement field should be in the form:

$$\underline{u}(r,\theta,\phi) = u(r,\theta,\phi).\underline{e}_r + v(r,\theta,\phi).\underline{e}_\theta + w(r,\theta,\phi).\underline{e}_\phi \qquad [10.169]$$

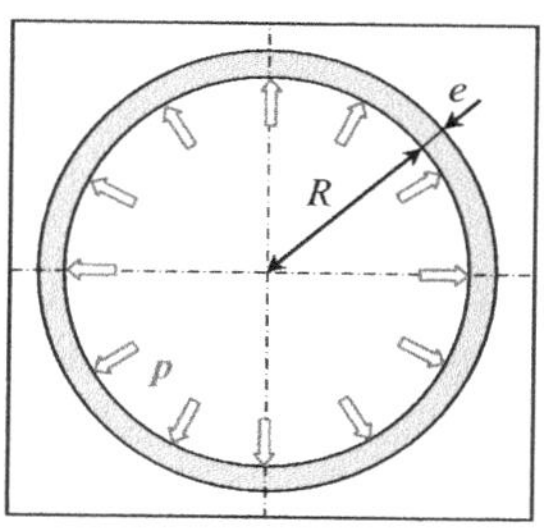

Figure 10.36. *Reservoir under pressure*

Since this problem is spherically symmetric, it is not a function of θ and ϕ (the displacement value must be the same irrespective of the value of θ and ϕ) and displacement is necessarily in the r-direction (if at one point we had displacement in the ϕ, since the plane (O,r,θ) is the problem's plane of symmetry, then this displacement is necessarily zero; the same applies for displacement in the θ-direction).

Hence:

$$\underline{u}(r,\theta,\phi) = u(r).\underline{e}_r \qquad [10.170]$$

QUESTION 2.–

The problem's boundary conditions are simply stress boundary conditions on the inside and outside of the sphere:

– in $r = R$: $\underline{\sigma}(M(r=R,\theta,\phi),\underline{n}_{ext}) = \underline{\underline{\sigma}}(M(R,\theta,\phi)).(-\underline{e}_r) = P.\underline{e}_r$, essentially, pressure inside the sphere is P and is in the r-direction.

– in $r = R + e$: $\underline{\sigma}(M(r=R+e,\theta,\phi),\underline{n}_{ext}) = \underline{\underline{\sigma}}(M(R,\theta,\phi)).\underline{e}_r = \underline{0}$, essentially, pressure outside the sphere is zero.

QUESTION 3.–

Navier's equation corresponds to an equilibrium equation in which we add the relationship between displacement and strain:

$$\underline{\underline{\varepsilon}} = \frac{1}{2}.\left(\underline{\underline{grad}}(\underline{u}) + \underline{\underline{grad}}^t(\underline{u})\right) \qquad [10.171]$$

and the behavior law between stress and strain, which for a linear elastic, homogeneous and isotropic material is here:

$$\underline{\underline{\sigma}} = 2.\mu.\underline{\underline{\varepsilon}} + \lambda.trace\left(\underline{\underline{\varepsilon}}\right).\underline{\underline{I}} \qquad [10.172]$$

Here, the only equation that is not zero is that in the r-direction (of course, remember to determine the Laplacian gradient or divergence operators in spherical coordinates):

$$(\lambda + 2.\mu).\left(u'' + \frac{2}{r}.u' - \frac{2}{r^2}.u\right) = 0 \qquad [10.173]$$

Since $\lambda + 2.\mu$ is not zero, we find a second-order differential equation. So, the solution is the linear combination of two linearly independent solutions. Here, we can search for solutions in the form r^n. And we find:

$$n.(n-1).r^{n-2} + \frac{2}{r}.n.r^{n-1} - \frac{2}{r^2}.r^n = 0 \qquad [10.174]$$

So we find that the characteristic polynomial solutions are $n = 1$ and $n = -2$, hence the solution:

$$u = A.r + \frac{B}{r^2} \qquad [10.175]$$

where A and B are two constants to be determined.

QUESTION 4.–

The strain tensor is:

$$\underline{\underline{\varepsilon}} = \frac{1}{2}.\left(\underline{\underline{grad}}(\underline{u}) + \underline{\underline{grad}}^t(\underline{u})\right) = \begin{bmatrix} u' & 0 & 0 \\ 0 & \frac{u}{r} & 0 \\ 0 & 0 & \frac{u}{r} \end{bmatrix}_{(\underline{r},\underline{\theta},\underline{\phi})}$$

$$= \begin{bmatrix} A - \frac{2.B}{r^3} & 0 & 0 \\ 0 & A + \frac{2.B}{r^3} & 0 \\ 0 & 0 & A + \frac{2.B}{r^3} \end{bmatrix}_{(\underline{r},\underline{\theta},\underline{\phi})} \qquad [10.176]$$

QUESTION 5.–

And the stress tensor:

$$\underline{\underline{\sigma}} = 2.\mu.\underline{\underline{\varepsilon}} + \lambda.trace\left(\underline{\underline{\varepsilon}}\right).\underline{\underline{I}}$$

$$= \begin{bmatrix} (3.\lambda + 2.\mu).A - \frac{4.\mu.B}{r^3} & 0 & 0 \\ 0 & (3.\lambda + 2.\mu).A + \frac{2.\mu.B}{r^3} & 0 \\ 0 & 0 & (3.\lambda + 2.\mu).A + \frac{2.\mu.B}{r^3} \end{bmatrix}_{(\underline{r},\underline{\theta},\underline{\phi})} \qquad [10.177]$$

QUESTION 6.–

The stress boundary conditions give us:

$$\begin{cases} -\sigma_r(R,\theta,\phi) = P \\ \sigma_r(R+e,\theta,\phi) = 0 \end{cases} \Rightarrow \begin{cases} (3.\lambda+2.\mu).A - \dfrac{4.\mu.B}{R^3} = -P \\ (3.\lambda+2.\mu).A - \dfrac{4.\mu.B}{(R+e)^3} = 0 \end{cases} \quad [10.178]$$

Hence:

$$\begin{cases} A = \dfrac{P}{3.\lambda+2.\mu}.\dfrac{R^3}{(R+e)^3 - R^3} \\ B = \dfrac{P}{4.\mu}.\dfrac{(R+e)^3.R^3}{(R+e)^3 - R^3} \end{cases} \quad [10.179]$$

QUESTION 7.–

The volume variation is:

$$\frac{\Delta V}{V} = tr\left(\underline{\underline{\varepsilon}}\right) = 3.A \quad [10.180]$$

Since $A > 0$, the volume variation is strictly positive, and volume increases.

QUESTION 8.–

If e/R is small compared with unity, we can then perform a first-order limited development (remember, $(1 + e/R)^n \approx 1 + n.e/R$). So, we obtain:

$$\begin{cases} \sigma_r \approx 0 \\ \sigma_\theta = \sigma_\phi \approx \dfrac{P.R}{2.e} \end{cases} \quad [10.181]$$

This is a very conventional result that you can also find by writing the equilibrium equation of a half sphere (see the last question).

QUESTION 9.–

If the material is ductile, then we can use the Tresca criterion (which is a criterion for the start of plasticity rather than a rupture criterion!):

$$\sigma_{tresca} = 2.\tau_{\max} = Max\left(\left|\sigma_I - \sigma_{II}\right|;\left|\sigma_{II} - \sigma_{III}\right|;\left|\sigma_{III} - \sigma_I\right|\right) < \sigma_e \qquad [10.182]$$

Since the stress tensor is diagonal here, we have the principal stress immediately. We obtain:

$$\sigma_{tresca} = 2.\tau_{\max} = \frac{P.R}{2.e} < \sigma_e \quad \Rightarrow \quad e_{alu} > \frac{P.R}{2.\sigma_e} = 17.1\,mm \qquad [10.183]$$

For a mass of:

$$M_{alu} = 4.\pi.R^2.e_{alu}.\rho_{alu} = 52\,kg \qquad [10.184]$$

QUESTION 10.–

A carbon/epoxy-type composite material is a brittle material. So, we use a maximum/minimum normal stress criterion:

$$\begin{cases} Max\left(\sigma_I, \sigma_{II}, \sigma_{III}\right) < \sigma_{trac} \\ Min\left(\sigma_I, \sigma_{II}, \sigma_{III}\right) > \sigma_{comp} \end{cases} \qquad [10.185]$$

Here, all of the stresses are positive, hence:

$$\frac{P.R}{2.e} < \sigma_{trac} \quad \Rightarrow \quad e_{tissu} > \frac{P.R}{2.\sigma_e} = 8.6\,mm \qquad [10.186]$$

For a mass of:

$$M_{tissu} = 4.\pi.R^2.e_{tissu}.\rho_{tissu} = 18\,kg \qquad [10.187]$$

Hence, a structure around three times lighter in carbon/epoxy composite fabric. In reality, this type of reservoir is indeed made of composite material.

QUESTION 11.–

In order to verify the hypothesis in question 8, we must calculate:

$$\frac{e_{tissu}}{R} = 2.9\% \qquad [10.188]$$

This ratio is less than one and the hypothesis is verified.

QUESTION 12.–

We have to write out the equilibrium of a half sphere. On the one hand, it is subjected to the force of pressure. So, you must integrate this pressure (the direction of which varies) or use the result stating that this integral is equal to the pressure multiplied by the projected surface, which here is $P.\pi.R^2$. On the other hand, it is subjected to the force due to the stress σ_θ (which we can assume to be constant since the thickness e is less than R), which here is $2.\pi.R.\sigma_\theta$. By equating the two terms, you will get the previous result.

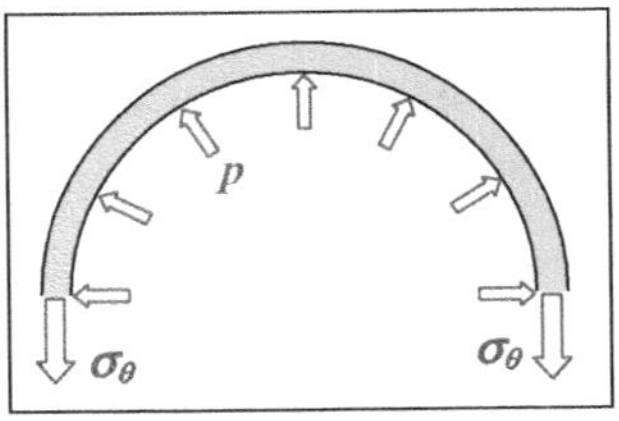

Figure 10.37. *Stress in a reservoir under pressure. For a color version of this figure, see www.iste.co.uk/bouvet/aeronautical.zip*

10.13. Plastic bending

QUESTION 1.–

If you recall the beam theory or using the results of exercise 17, we can show that when bending:

$$\sigma_x = \frac{-M_{fz}}{I_z}.y \qquad [10.189]$$

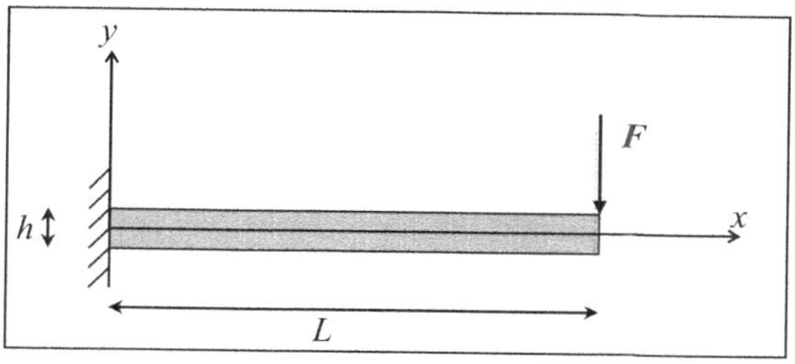

Figure 10.38. *Plastic bending*

where M_{fz} is the bending moment, here:

$$M_{fz}(x) = -F.(L-x) \qquad [10.190]$$

and I_z the quadratic moment of inertia of the cross-section (here rectangular):

$$I_z = \iint_S y^2.dS = \frac{b.h^3}{12} \qquad [10.191]$$

We obtain the classic result of a linear stress function of x and y, which is maximum at the clamping point:

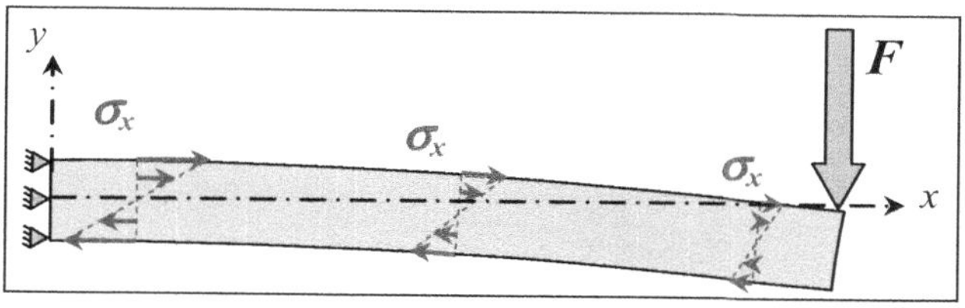

Figure 10.39. *Stress in a beam under elastic bending. For a color version of this figure, see www.iste.co.uk/bouvet/aeronautical.zip*

We should also show that there is a quadratic shear stress τ_{xy}, null at the top and bottom of the cross-section, which is negligible compared to the longitudinal stress σ_x in the case of a full cross-section (see exercise 17). This shear stress is to be disregarded in this exercise.

Clearly, the strain field is in the same form as the stress field *($\varepsilon_x = \sigma_x/E$)*.

You can also find the result for the stress field by writing that the integral of σ_x on a cross-section is equal to the cohesive force. Cohesive force is of course made up of a shear force $T_y = F$ and a bending moment $M_{fz} = -F.(L-x)$. If you integrate the moment created by σ_x in a cross-section, you will find the $K(x)$ expression directly. And if you integrate the force created by σ_x on a cross-section, then you will find zero (it is easy to see graphically in the figure above that the sum of all the red arrows is zero)! In order to find the shear force, we must integrate τ_{xy} which is quadratic, but is neglected here. This point is treated in detail in exercise 17 concerning the bending beam.

QUESTION 2.–

Plasticity will appear at the place where stress is maximum. The term stress is clearly insufficient, as typically it concerns a 3D stress tensor. Since the material is ductile (aluminum), here we assume that it follows a Von Mises criterion:

$$\sigma_{VM} < \sigma_e \qquad [10.192]$$

Here, the Von Mises criterion is equal (in absolute value) to σ_x:

$$\sigma_{VM} = \sqrt{\frac{1}{2}.\left((\sigma_I - \sigma_{II})^2 + (\sigma_{II} - \sigma_{III})^2 + (\sigma_{III} - \sigma_I)^2\right)} = |\sigma_x| \qquad [10.193]$$

Plasticity will begin at $x = 0$ and $y = \pm h/2$, for a force F_e such that:

$$|\sigma_x| = \frac{12.F_e.L}{b.h^3}.\frac{h}{2} = \sigma_e \quad \Rightarrow \quad F_e = \frac{\sigma_e.b.h^2}{6.L} \qquad [10.194]$$

QUESTION 3.–

Once plasticity has occurred, it spreads through the beam from these points. We cannot use the distribution of linear stress function of y to determine these zones, as we did earlier, since the stress will saturate at $\sigma_x = \sigma_e$, while strain ε_x will remain linear with y at the same time:

In order to determine the plasticity zone, we have to write that the bending moment (which is always equal to $-F.(L-x)$) is equal to the integral of the moment created by this stress distribution (not neglecting to write that σ_x is linear at the center then equal to σ_e):

$$M_{fz} = -F.(L-x) = -\iint_S \sigma_x.y.dS = -2.b.\left[\int_0^{\delta} \frac{\sigma_e}{\delta}.y^2.dy + \int_{\delta}^{h/2} \sigma_e.y.dy\right] \qquad [10.195]$$

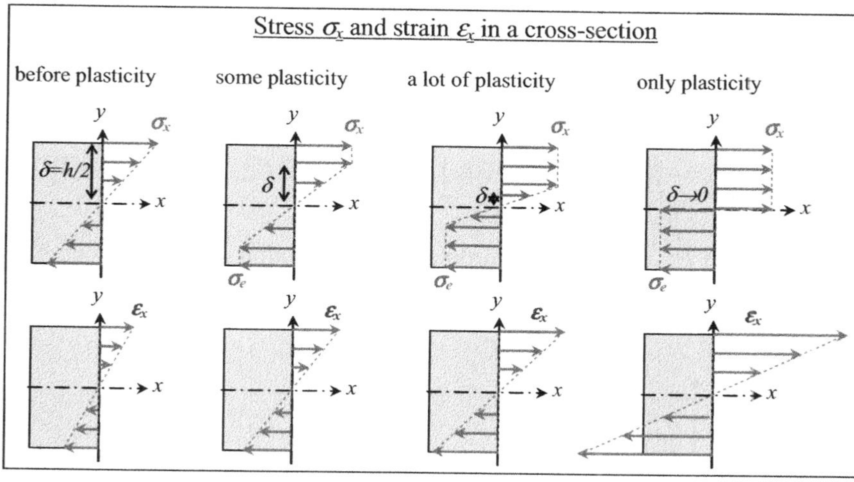

Figure 10.40. *Stress and strain in an elastic–plastic bending beam. For a color version of this figure, see www.iste.co.uk/bouvet/aeronautical.zip*

In this way we can determine δ that depends on x and F, then plot the plastic zone's development:

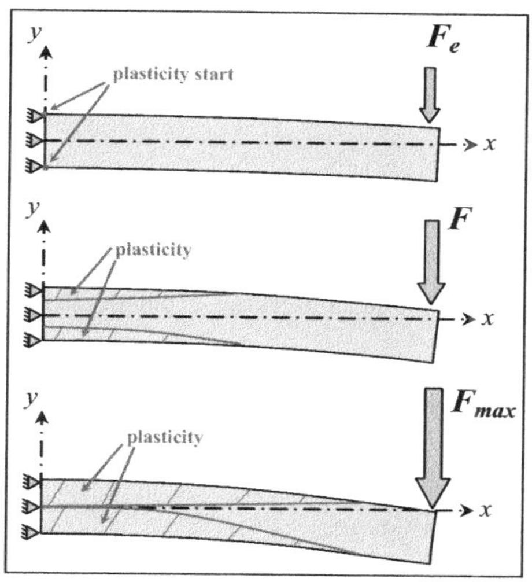

Figure 10.41. *Plastic zone in an elastic–plastic bending beam. For a color version of this figure, see www.iste.co.uk/bouvet/aeronautical.zip*

We can then calculate the maximum force F_{max} that can possibly be applied to the beam. This maximum force corresponds to cases in which the whole cross-section of the beam (here in $x = 0$) is plastic ($\delta \to 0$):

$$M_{fz\max} = -F_{\max}.L = -\iint_S \pm\sigma_e.y.dS = -2.b.\int_0^{h/2} \sigma_e.y.dy \quad \Rightarrow \quad F_{\max} = \frac{\sigma_e.b.h^2}{4.L} \qquad [10.196]$$

So, we obtain $F_{max} = 1.5.F_e$. According to the deflection d, the force curve will be linear up to F_e, then becomes nonlinear, and becoming asymptotic towards F_{max}:

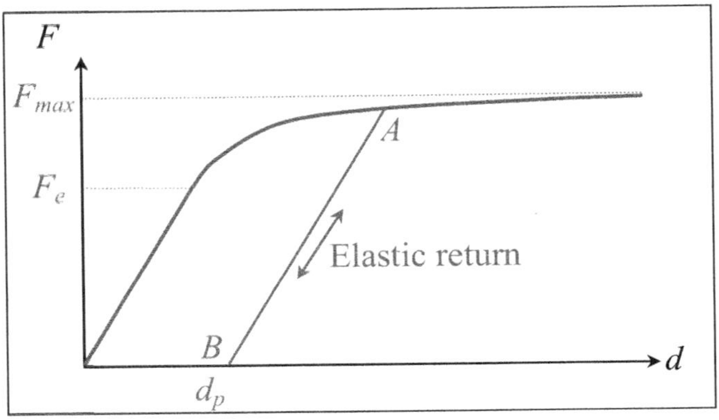

Figure 10.42. *Plastic behavior of an elastic–plastic bending beam. For a color version of this figure, see www.iste.co.uk/bouvet/aeronautical.zip*

QUESTION 4.–

If we release the force at point A having reached plasticity, we discharge along the same slope as the initial slope, observing a plastic strain of the beam d_p (see previous figure).

So, having released force F, we obtain a stress field in the cross-section with a zero:

$$M_{fz}(B) = 0 = -\iint_S \sigma_x.y.dS \qquad [10.197]$$

That is a stress field in the form:

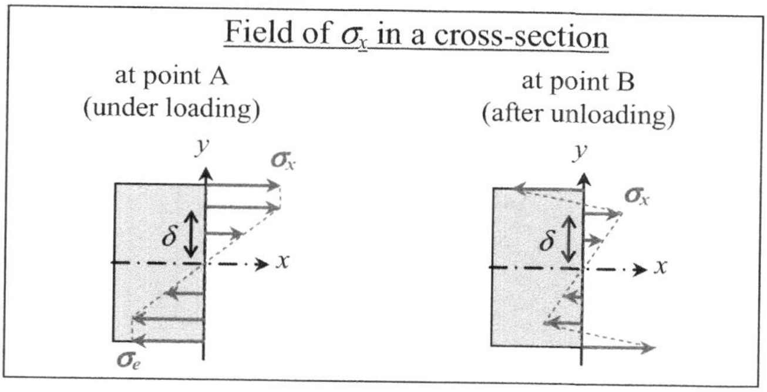

Figure 10.43. *Stress in an elastic–plastic bending beam. For a color version of this figure, see www.iste.co.uk/bouvet/aeronautical.zip*

Indeed, we see that the most loaded points (at $y = \pm h/2$) are those with the greatest plastic strain and will consequently generate negative stress when the loading is released.

10.14. Disc under radial tension

Part 1: conventional method

QUESTION 1.–

Since the disc is thin, we can consider it as being the object of plane stress (while a thick disc would be the object of plane strain instead). Moreover, the problem presents a central symmetry at center O, so $\tau_{r\theta} = 0$, hence:

$$\underline{\underline{\sigma}} = \begin{bmatrix} \sigma_r & 0 & 0 \\ 0 & \sigma_\theta & 0 \\ 0 & 0 & 0 \end{bmatrix}_{(\underline{r},\underline{\theta},\underline{z})} \quad [10.198]$$

So, we have strain in the form:

$$\underline{\underline{\varepsilon}} = \frac{1+\nu}{E}\underline{\underline{\sigma}} - \frac{\nu}{E}.trace\left(\underline{\underline{\sigma}}\right).\underline{\underline{I}} = \frac{1}{E}.\begin{bmatrix} \sigma_r - \nu.\sigma_\theta & 0 & 0 \\ 0 & \sigma_\theta - \nu.\sigma_r & 0 \\ 0 & 0 & -\nu(\sigma_r + \sigma_\theta) \end{bmatrix}_{(\underline{r},\underline{\theta},\underline{z})} \quad [10.199]$$

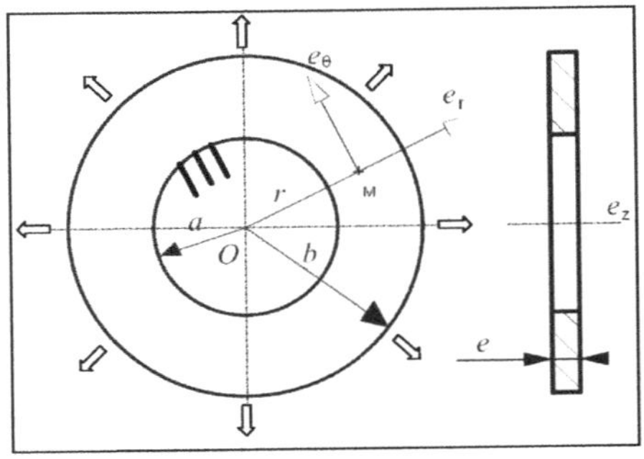

Figure 10.44. *Disc under radial tension*

Remember that we cannot have plane stress and plane strain at the same time.

QUESTION 2.–

A priori the displacement field should have the form:

$$\underline{u}(r,\theta,z) = u(r,\theta,z).\underline{e}_r + v(r,\theta,z).\underline{e}_\theta + w(r,\theta,z).\underline{e}_z \quad [10.200]$$

But since the problem is centrally symmetric, it is not function of θ (the displacement value must be the same irrespective of the value of θ) and displacement in the θ-direction is necessarily zero (if at one point we had displacement in the θ-direction, since the plane *(O,r,z)* is the problem's plane of symmetry, then this displacement would necessarily be zero).

Hence:

$$\underline{u}(r,\theta,z) = u(r,z).\underline{e}_r + w(r,z).\underline{e}_z \quad [10.201]$$

The last conditions are the most difficult to justify (u is not function of z and w is not function of r), and moreover we will see below that there are only approximations.

If we assume stress is homogeneous through the thickness, then so is strain. So, displacement u will also not be function of z. The fact that we then assume that $\varepsilon_{rz} = 0$ necessarily implies w is not function of r (I leave you to write the displacement gradient) and we find $\partial w / \partial r = 0$, so we can assume:

$$\underline{u}(r,\theta,z) = u(r).\underline{e}_r + w(z).\underline{e}_z \quad [10.202]$$

And using the Navier's equation:

$$\mu.\underline{\Delta}(\underline{u})+(\lambda+\mu).\underline{grad}\left(div(\underline{u})\right)+\underline{f}_v=\underline{0} \qquad [10.203]$$

We then find:

$$\begin{cases} \dfrac{\partial^2 \mathrm{u}}{\partial r^2}+\dfrac{1}{r}.\dfrac{\partial \mathrm{u}}{\partial r}-\dfrac{u}{r^2}=0 \\ \dfrac{\partial^2 \mathrm{w}}{\partial z^2}=0 \end{cases} \qquad [10.204]$$

which then provides the resolution:

$$\begin{cases} u=A.r+\dfrac{B}{r} \\ w=C+D.z \end{cases} \qquad [10.205]$$

where A, B, C and D are the integration constants.

QUESTION 3.–

We must then write the boundary conditions in order to determine the integration constants. The displacement boundary conditions require a radial displacement of zero in the interior radius and a z-displacement of zero on one side of the disc (for example in $z = 0$):

$$\begin{cases} u(r=a)=0 \\ w(z=0)=0 \end{cases} \qquad [10.206]$$

In this way you determine B and C, hence:

$$\begin{cases} u=A.\left(r-\dfrac{a^2}{r}\right) \\ w=D.z \end{cases} \qquad [10.207]$$

We can now determine strain:

$$\underline{\underline{\varepsilon}} = \frac{1}{2}.\left(\underline{\underline{grad}}(\underline{u}) + \underline{\underline{grad}}^t(\underline{u})\right) = \begin{bmatrix} A.\left(1+\dfrac{a^2}{r^2}\right) & 0 & 0 \\ 0 & A.\left(1-\dfrac{a^2}{r^2}\right) & 0 \\ 0 & 0 & D \end{bmatrix}_{(\underline{r},\underline{\theta},\underline{z})} \quad [10.208]$$

In addition, to respect the plane stress hypothesis, we must have:

$$\sigma_z = \lambda.(\varepsilon_r + \varepsilon_\theta) + (\lambda + 2.\mu)..\varepsilon_z = 0 \quad [10.209]$$

Hence:

$$D = \frac{-\lambda}{\lambda + 2.\mu}.A \quad [10.210]$$

Now for stress:

$$\underline{\underline{\sigma}} = 2.\mu.\underline{\underline{\varepsilon}} + \lambda.trace\left(\underline{\underline{\varepsilon}}\right).\underline{\underline{I}}$$
$$= \begin{bmatrix} 2.A.\mu.\left(1+\dfrac{a^2}{r^2}+\dfrac{2.\lambda}{\lambda+2.\mu}\right) & 0 & 0 \\ 0 & 2.A.\mu.\left(1-\dfrac{a^2}{r^2}+\dfrac{2.\lambda}{\lambda+2.\mu}\right) & 0 \\ 0 & 0 & 0 \end{bmatrix}_{(\underline{r},\underline{\theta},\underline{z})} \quad [10.211]$$

Now we must write the stress boundary condition in $r = b$ to determine the A constant:

– in $r = b$: $\underline{\sigma}(M(r=b,\theta,z),\underline{n}_{ext}) = \underline{\underline{\sigma}}(M(b,\theta,z)).\underline{e}_r = P.\underline{e}_r$ [10.212]

hence:

$$\sigma_r(b) = P \quad \Rightarrow \quad A = \frac{P}{2.\mu\left(\dfrac{3.\lambda+2.\mu}{\lambda+2.\mu} + \dfrac{a^2}{b^2}\right)} \qquad [10.213]$$

So, after numerical application:

$$\begin{cases} u(b) = 6.35.10^{-2}\ mm \\ \sigma_r(a) = 149\ MPa \\ \sigma_r(b) = 100\ MPa \\ \sigma_\theta(a) = 44.7\ MPa \\ \sigma_\theta(b) = 93.6\ MPa \end{cases} \qquad [10.214]$$

Of course, you can see that the $\sigma_r(b) = P$ boundary condition is indeed verified.

The state of maximum stress (meaning the maximum normal stress, i.e. in cases where the material is assumed to be brittle) is reached in $r = a$ for a facet of normal vector $\underline{e}r$; so if a crack should develop, it would be in the plane (θ, z).

Part 2: energy-based method

QUESTION 4.–

The easiest choice of displacement field is to take a linear field function of r:

$$\underline{u}(r,\theta,z) = A'.(r-a).\underline{e}_r \qquad [10.215]$$

This choice allows us to respect the null displacement in $r = a$.

Remember that the Ritz method allows us to determine the best (in terms of the difference between strain energy and work of external forces, which would be zero if we had the exact solution) function of those proposed. So if your proposed functions are pertinent, you will find a pertinent solution, otherwise you will only find the "least ill-adapted" of the functions you proposed!

In this choice of displacement field, we can see that stress is not plane:

$$\underline{\underline{\varepsilon}} = \begin{bmatrix} A' & 0 & 0 \\ 0 & A'\left(1-\dfrac{a}{r}\right) & 0 \\ 0 & 0 & 0 \end{bmatrix}_{(\underline{r},\underline{\theta},\underline{z})} \qquad [10.216]$$

Indeed:

$$\underline{\underline{\sigma}} = \begin{bmatrix} A'.\left(2.(\lambda+\mu)-\lambda.\dfrac{a}{r}\right) & 0 & 0 \\ 0 & A'.\left(2.(\lambda+\mu)-(\lambda+2.\mu).\dfrac{a}{r}\right) & 0 \\ 0 & 0 & A'.\lambda.\left(2.-\dfrac{a}{r}\right) \end{bmatrix}_{(\underline{r},\underline{\theta},\underline{z})} \qquad [10.217]$$

For them to be plane, we must make a broader choice of displacement field, for example:

$$\underline{u}(r,\theta,z) = A'.(r-a).\underline{e}_r + B'.z.\underline{e}_z \qquad [10.218]$$

And by imposing a condition on B', we would have plane stress.

QUESTION 5.–

We must then determine the strain energy:

$$E_d = \frac{1}{2}.\iiint_V \underline{\underline{\sigma}} : \underline{\underline{\varepsilon}}.dV = \frac{1}{2}.2\pi.e.\int_{r=a}^{b} \left(\sigma_x.\varepsilon_x + \sigma_y.\varepsilon_y\right).dr \qquad [10.219]$$

This rather trying calculation gives us:

$$E_d = \pi.A'^2.e.\left(4.(\lambda+\mu).\left(\frac{b^2-a^2}{2} - a.(b-a\right) + (\lambda+2.\mu).a^2.\ln\left(\frac{b}{a}\right)\right) \qquad [10.220]$$

QUESTION 6.–

We must then calculate the work of the external forces. Here, this work is created in $r = b$ throughout the disc periphery (by applying pressure P, the displacement $u(b)$ is induced):

$$W_{ext} = \frac{1}{2}\cdot\iint_{Sb} P.u(b).dS = \frac{1}{2}.2\pi.e.P.A'.(b-a) \quad [10.221]$$

QUESTION 7.–

Now we just write that $(E_d - W_{ext})$ is minimum for the best value of A', so:

$$\frac{\partial(E_d - W_{ext})}{\partial A'} = 0 \quad [10.222]$$

which after the calculation gives:

$$A' = \frac{P.b.(b-a)}{2.(\lambda+\mu).(a-b)^2 + (\lambda+2.\mu).a^2.(\ln(b)-\ln(a))} \quad [10.223]$$

QUESTION 8.–

Then we obtain the following numerical calculation:

$$\begin{cases} u(b) = 4.69.10^{-2}\ mm \\ \sigma_r(a) = 84.2\ MPa \\ \sigma_r(b) = 111\ MPa \\ \sigma_\theta(a) = 36.1\ MPa \\ \sigma_\theta(b) = 99.3\ MPa \end{cases} \quad [10.224]$$

Note that the $\sigma_r(b) = P$ boundary condition is not verified. This is typical for an energy-based method; the only boundary conditions respected are those imposed in the choice of the field imposed (here the chosen displacement field imposes $u(a) = 0$).

We see a significant difference in the displacement field: $6.35.10^{-2}$ mm for the conventional method compared to $4.69.10^{-2}$ mm for the energy-based method.

In reality, much of this difference is due to the hypothesis of plane stress. Indeed, if you repeated the first part of this exercise choosing the hypothesis of plane strain (hypothesis adapted for thick disk), you would have found:

$$u(r) = A''.\left(r - \frac{a^2}{r}\right) \qquad [10.225]$$

And:

$$A'' = \frac{P}{2.\left(\lambda + \mu.\left(1 + \frac{a^2}{b^2}\right)\right)} \qquad [10.226]$$

Hence, there is a displacement in $r = b$ of 0.047 mm, which is in keeping with the energy-based method.

10.15. Bending beam: resolution by the Ritz method

QUESTION 1.–

To ensure that the solution reached using the Ritz method is pertinent, we must respect the boundary conditions as much as possible. Here, we must have null displacement in $x = 0$ and in $x = L$, hence the displacement field (choosing a sinusoidal function) of the beam's neutral fiber:

$$v_0(x) = \delta.\sin\left(\frac{\pi.x}{L}\right) \qquad [10.227]$$

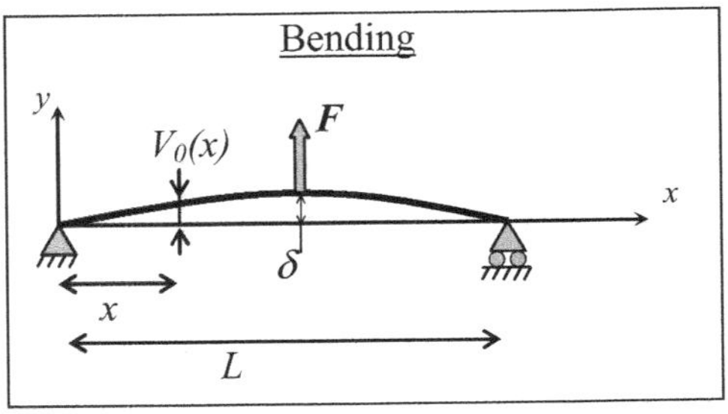

Figure 10.45. *Simply supported beam in bending*

Draw this function to convince yourself that the form is indeed coherent with the required form.

QUESTION 2.–

The previous function allows us to determine the neutral fiber's displacement. Of course, we also need the full displacement field in order to determine the strain field, then the stress field, before applying the Ritz method. To this end, we have to assume that a straight cross-section remains straight and perpendicular to the neutral fiber after deformation. We should also note that the rotation angle of a cross-section is, at the first order, equal to the derivation of function v_0.

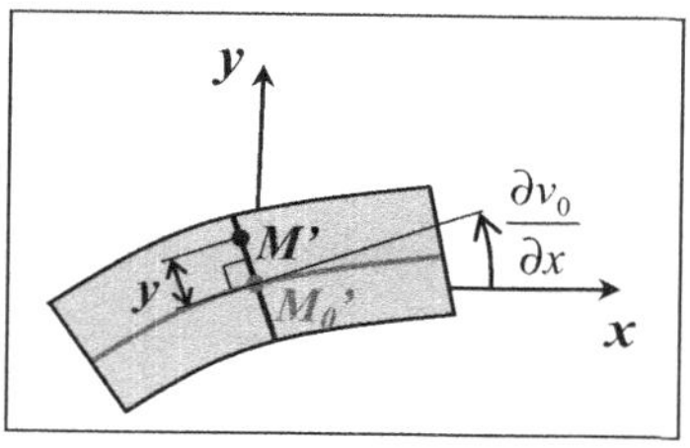

Figure 10.46. *Kinematic of a bending beam. For a color version of this figure, see www.iste.co.uk/bouvet/aeronautical.zip*

Hence:

$$\begin{cases} u(x,y) \approx \dfrac{-\partial v_0}{\partial x}.y = -\delta.\dfrac{\pi}{L}.\cos\left(\dfrac{\pi.x}{L}\right) \\ v(x,y) \approx v_0(x) = \delta.\sin\left(\dfrac{\pi.x}{L}\right) \end{cases} \quad [10.228]$$

You can also note that if we assume a small angle of rotation, the displacement in the y-direction at all points of a cross-section is assumed to be the same.

QUESTION 3.–

The strain field is then:

$$\underline{\underline{\varepsilon}} = \frac{1}{2}.\left(\underline{\underline{grad}}(\underline{u}) + \underline{\underline{grad}}^t(\underline{u})\right) = \begin{bmatrix} \delta.\left(\dfrac{\pi}{L}\right)^2.y.\sin\left(\dfrac{\pi.x}{L}\right) & 0 \\ 0 & 0 \end{bmatrix}_{(\underline{x},\underline{y})} \quad [10.229]$$

And at a first approximation, the bending only generates a strain ε_x. We will consider this again for the next question.

QUESTION 4.–

The stress field is then:

$$\underline{\underline{\sigma}} = 2.\mu.\underline{\underline{\varepsilon}} + \lambda.trace\left(\underline{\underline{\varepsilon}}\right).\underline{\underline{I}}$$

$$= \begin{bmatrix} (\lambda+2.\mu).\delta.\left(\dfrac{\pi}{L}\right)^2.y.\sin\left(\dfrac{\pi.x}{L}\right) & 0 \\ 0 & \lambda.\delta.\left(\dfrac{\pi}{L}\right)^2.y.\sin\left(\dfrac{\pi.x}{L}\right) \end{bmatrix}_{(\underline{x},\underline{y})} \quad [10.230]$$

So, we obtain a σ_x stress field conforming to a bending stress field:

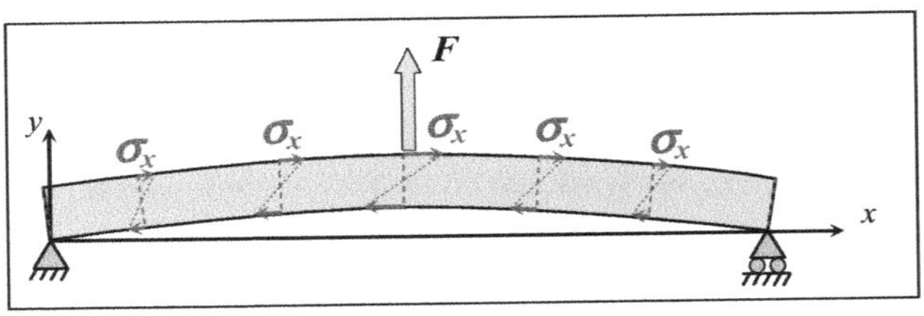

Figure 10.47. *Stress in a simply supported bending beam. For a color version of this figure, see www.iste.co.uk/bouvet/aeronautical.zip*

On the other hand, the stress field σ_y is wrong (so I will not ask you to plot it). We also find a null τ_{xy} shear stress field, which, in reality, is not the case. Indeed, we can show that the equilibrium equation imposes a parabolic τ_{xy} field, null on top and under the beam and maximum at the center (see exercise 17 for more details).

QUESTION 5.–

We can then determine the strain energy:

$$E_d = \frac{1}{2}.\iiint_V \underline{\underline{\sigma}} : \underline{\underline{\varepsilon}}.dV = \frac{1}{2}.\int_{x=0}^{L}\int_S \sigma_x.\varepsilon_x.dS.dx = \frac{1}{4}.(\lambda+2.\mu).\delta^2.\frac{\pi^4}{L^3}.I_z \quad [10.231]$$

where I_z is the quadratic moment of inertia of the cross-section:

$$I_z = \iint_S y^2.dS \qquad [10.232]$$

Next, we must determine the work of the external forces. Here supporting forces do not work (as there is null displacement) and only the force in the beam center works:

$$W_{ext} = \frac{1}{2}.F.v_0(x = L/2) = \frac{1}{2}.F.\delta \qquad [10.233]$$

Now we only have to write that $(E_d - W_{ext})$ is minimum for the best value of δ, so:

$$\frac{\partial(E_d - W_{ext})}{\partial\delta} = 0 \qquad [10.234]$$

which gives:

$$\delta = \frac{2.F.L^3}{\pi^4.(\lambda + 2.\mu).I_z} \qquad [10.235]$$

We can compare this result to the exact result (that you can show by recalling the lesson on beams or by reading [AGA 08] for instance):

$$\delta = \frac{F.L^3}{48.E.I_z} \qquad [10.236]$$

Here, noting that:

$$\lambda + 2.\mu = \frac{(1-\nu).E}{(1+\nu).(1-2.\nu)} \qquad [10.237]$$

And you find yourself comparing:

$$\frac{\pi^4.(1-\nu)}{2.(1+\nu).(1-2.\nu)} \quad \text{and} \quad 48 \qquad [10.238]$$

If $\nu = 0$, the first coefficient is 48.7, and the result of the energy-based method is of very good quality. On the other hand, if $\nu = 0.3$, the coefficient is 65.6 and the result is incorrect! Indeed, here it is not the energy-based method that is in error but the hypothesis on the form of the stress/strain tensor. The hypothesis used for the displacement field requires that the only non-zero strain tensor components ε_x and consequently the three diagonal terms of the stress tensor are non-zero. In reality, we see the opposite: only the σ_x term of the stress tensor is non-zero, and the three diagonal terms of the strain tensor stem from this. Had we taken this hypothesis, we would have found $\sigma_x = E.\varepsilon_x$ and so:

$$\delta = \frac{2.F.L^3}{\pi^4.E.I_z} \qquad [10.239]$$

which is a very good result compared to the exact beam solution (or nearly exact, as even this beam solution is also an approximate solution).

10.16. Stress concentration in open hole

So:

$$\underline{\underline{\sigma}}(M) = \begin{bmatrix} \sigma_r & \tau_{r\theta} & 0 \\ \tau_{r\theta} & \sigma_\theta & 0 \\ 0 & 0 & 0 \end{bmatrix}_{(\underline{r},\underline{\theta},\underline{z})} \qquad [10.240]$$

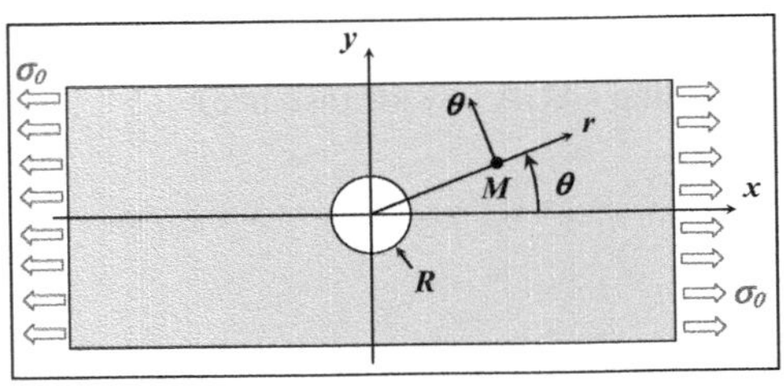

Figure 10.48. *Open hole tension test. For a color version of this figure, see www.iste.co.uk/bouvet/aeronautical.zip*

And:

$$\begin{cases} \dfrac{\sigma_r}{\sigma_0} = \dfrac{1}{2}.\left[1 - \dfrac{1}{\rho^2} + \left(1 - \dfrac{4}{\rho^2} + \dfrac{3}{\rho^4}\right).\cos(2.\theta)\right] \\ \dfrac{\sigma_\theta}{\sigma_0} = \dfrac{1}{2}.\left[1 + \dfrac{1}{\rho^2} - \left(1 + \dfrac{3}{\rho^4}\right).\cos(2.\theta)\right] \\ \dfrac{\tau_{r\theta}}{\sigma_0} = -\dfrac{1}{2}.\left(1 + \dfrac{2}{\rho^2} - \dfrac{3}{\rho^4}\right).\sin(2.\theta) \end{cases} \quad \text{with } \rho = \frac{r}{R} \qquad [10.241]$$

QUESTION 1.–

To show that this stress field is the solution here, we have to show that it verifies all of the problem's conditions such as:

– The equilibrium equation. You can indeed show that:

$$\underline{div}\left(\underline{\underline{\sigma}}\right)=\underline{0} \qquad [10.242]$$

Since the volume forces are zero. Do not forget to write the divergence in cylindrical coordinates!

– The stress boundary conditions:

- At the hole, that is, in $\rho = 1$, the external forces are zero, hence:

$$\underline{\sigma}(M(\rho=1),\underline{n}_{ext})=\underline{\underline{\sigma}}(M(\rho=1)).(-\underline{e}_r)=\underline{0} \qquad [10.243]$$

You can therefore show that this relationship is true.

- At infinity, that is, when $\rho \to +\infty$, the external forces are zero for a facet of normal vector $\underline{y}$ and equal to $\sigma_0.\underline{x}$ for a facet of normal vector $\underline{x}$:

$$\begin{cases}\underline{\sigma}(M(\rho\to+\infty),\underline{x})=\sigma_0.\underline{x}\\ \underline{\sigma}(M(\rho\to+\infty),\underline{y})=\underline{0}\end{cases} \qquad [10.244]$$

You can show that these relationships are indeed verified. You can perform the calculations in cylindrical or Cartesian coordinates, but when you compare or multiply the magnitudes (vectors or tensors); they must all be in the same coordinate system!

For example, you can start by showing that:

$$\underline{\underline{\sigma}}\underset{\rho\to+\infty}{\to}\begin{bmatrix}\sigma_0.\cos^2\theta & -\sigma_0.\cos\theta.\sin\theta & 0\\ -\sigma_0.\cos\theta.\sin\theta & \sigma_0.\sin^2\theta & 0\\ 0 & 0 & 0\end{bmatrix}_{(\underline{r},\underline{\theta},\underline{z})}=\begin{bmatrix}\sigma_0 & 0 & 0\\ 0 & 0 & 0\\ 0 & 0 & 0\end{bmatrix}_{(\underline{x},\underline{y},\underline{z})} \qquad [10.245]$$

We can then determine the strain field, and then verify that the strain field does verify the compatibility condition. Then we can finally integrate this strain field to determine the displacement field. In practice, we can show that the compatibility condition is not totally satisfied and that the proposed solution is not the exact solution (but that it is a good approximation). Since this compatibility condition is not totally satisfied, we cannot integrate the strain field to determine the displacement field; at least not in 3D (although in 2D, we can show that it is possible).

In conclusion, the proposed stress field is a good approximation of the solution, but it is not the exact solution here.

QUESTION 2.–

The only component of the non-zero stress in $r = R$ is σ_θ:

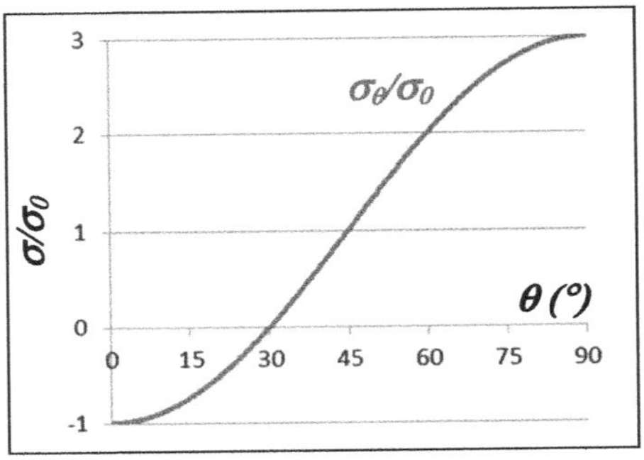

Figure 10.49. *Stress field around a hole. For a color version of this figure, see www.iste.co.uk/bouvet/aeronautical.zip*

QUESTION 3.–

The stress concentration factor at the edge of the hole is clearly reached at the edge of the hole, i.e. at $r = R$, and is:

– $K_t = 3$ in tension:

This is a conventional result worth retaining: when you make a hole of small size compared with the surrounding structure (for example, located far from surrounding holes, which is seldom the case), then the stress factor at the edge is 3! In other words, the fact that the hole is small divides the structure's strength by 3. Indeed, this result is true for a brittle material but not for a ductile material. For the ductile material, the stress concentration at the hole edge leads to localized plasticity that smooths the stress distribution at the hole edge, thereby decreasing the stress concentration (contrary to the received notion that it must be avoided at any cost, plasticity can in fact have a positive effect on a structure's strength). We can even show that this result does not apply to brittle materials that can damage themselves, such as the composite material used in aeronautical structures (a unidirectional laminate, for instance, with carbon fibers and epoxy resin). In this case, the damage (in terms of matrix cracks, delamination and localized fiber failure) plays the same role as plasticity and decreases the stress concentration due to damage (contrary to the received notion that it must be avoided at any cost, damage can in fact have a positive effect on a structure's strength).

Furthermore, if we reach structural failure, since σ_θ is the only non-zero stress the crack propagates in the *(r, z)* plane, beginning at $r = R$ and $\theta = \pi/2$, so at the top of the hole ($x = 0$ and $y = R$) as we could well expect.

– $K_t = 1$ in compression:

In fact, the curve above is more complex than it appears! We observe a stress concentration of 3. This result is true in tension as well as in compression. In other words, if you apply tension, you will see a tension stress three times greater than the imposed tension, and if you apply compression, you will see a compression stress three times greater than the imposed compression. But at the same time, you will see compression of the same value as the stress applied (in absolute value and in $r = R$ and $\theta = 0$) during a tension test, and tension of the same value as the applied stress (in absolute value and in $r = R$ and $\theta = 0$) during a compression test. This is a much less intuitive result than the previous one.

So, if you are working with a ductile material for which the elasticity limits for tension and compression are of the same order, then the start of plasticity will be at $r = R$ and $\theta = \pi/2$ and with a K_t of 3, both in tension and in compression. But if you are working with a brittle material for with the failure limit under tension is much less than the compression limit (in absolute value), then in tension you will reach failure at $r = R$ and $\theta = \pi/2$ with a K_t of 3, but in compression you will reach failure at $r = R$ and $\theta = 0$ with a K_t of 1!

10.17. Bending beam

Part 1: stress tensor

QUESTION 1.–

We must isolate the right part of the beam:

$$\begin{Bmatrix} \underline{F} \\ \underline{M}(G) \end{Bmatrix}_{ext/r} + \begin{Bmatrix} \underline{F} \\ \underline{M}(G) \end{Bmatrix}_{l/r} = \begin{Bmatrix} \underline{0} \\ \underline{0} \end{Bmatrix} \quad [10.246]$$

where G is the center of gravity for the cross-section *S(x)* in question. Hence the torsor of the internal forces of the right part on the left part is:

$$\begin{Bmatrix} \underline{F} \\ \underline{M}(G) \end{Bmatrix}_{r/l} = \begin{Bmatrix} \underline{F} \\ \underline{M}(G) \end{Bmatrix}_{ext/r} \quad [10.247]$$

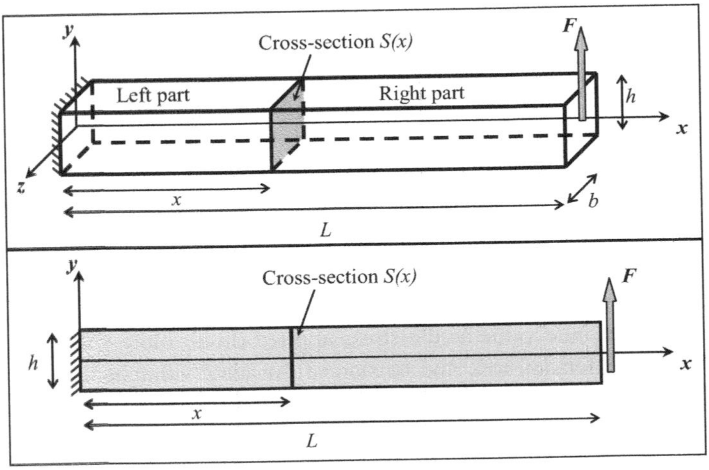

Figure 10.50. *Bending beam*

And the only force on the right part is the force F, which we must move to G, hence:

$$\left\{\begin{matrix}\underline{F}\\\underline{M}(G)\end{matrix}\right\}_{r/l}=\left\{\begin{matrix}T_y.\underline{y}\\M_z(G).\underline{z}\end{matrix}\right\}=\left\{\begin{matrix}F.\underline{y}\\F.(L-x).\underline{z}\end{matrix}\right\} \quad [10.248]$$

The shear force T_y will induce τ_{xy} stress (stress in the y-direction for a facet of normal vector $\underline{x}$) and the bending moment M_z will induce the normal stress σ_x (stress in the x-direction for a facet of normal vector $\underline{x}$).

QUESTION 2.–

The stress tensor has to verify the equilibrium equation (here without volume force):

$$\underline{div}\left(\underline{\underline{\sigma}}\right)+\underline{f}_v=\underline{div}\left(\underline{\underline{\sigma}}\right)=\underline{0} \quad [10.249]$$

It must also verify the compatibility condition written in stress via the behavior law (to ensure that once the strain tensor is determined, it can be integrated to determine displacement):

$$\underline{\underline{\Delta\sigma}}+\frac{1}{1+\nu}.\underline{\underline{grad}}\left(\underline{grad}\left(trace\left(\underline{\underline{\sigma}}\right)\right)\right)=\frac{-\nu}{1-\nu}.div\left(\underline{f}_v\right).\underline{\underline{I}}-\left(\underline{\underline{grad}}^t\left(\underline{f}_v\right)+\underline{\underline{grad}}\left(\underline{f}_v\right)\right) \quad [10.250]$$

It must also verify the stress boundary conditions (here written in 2D):

– on the top side: $\underline{\underline{\sigma}}\left(M(x,h/2),\underline{y}\right)=\underline{0}$;

– on the bottom side: $\underline{\underline{\sigma}}\left(M(x,-h/2),-\underline{y}\right)=\underline{0}$;

– on the right side of the beam: $\begin{cases} \iint_{SL}\underline{\underline{\sigma}}\left(M(L,y),\underline{x}\right).dS=F.\underline{y} \\ \iint_{SL}\underline{GM}\wedge\underline{\underline{\sigma}}\left(M(L,y),\underline{x}\right).dS=\underline{0} \end{cases}$.

where *G (L, 0)* is the center of gravity of the right side of the beam. Of course, we cannot assume that stress is homogeneous on this side, indeed, below we will show that this is not the case!

It must also verify the displacement boundary condition. Essentially, it has to determine the strain tensor by means of the behavior law:

$$\underline{\underline{\varepsilon}}=\frac{1+\nu}{E}\underline{\underline{\sigma}}-\frac{\nu}{E}.trace\left(\underline{\underline{\sigma}}\right).\underline{\underline{I}} \qquad [10.251]$$

then integrate to determine the displacements:

$$\underline{\underline{\varepsilon}}=\frac{1}{2}.\left(\underline{\underline{grad}}(\underline{u})+\underline{\underline{grad}}^{t}(\underline{u})\right) \qquad [10.252]$$

and verify that the displacement field verifies:

$$\underline{u}(M(0,y))=\underline{0} \qquad [10.253]$$

Considering the hypotheses made for the form of the stress tensor, we can then verify that this boundary condition is not totally verified!

QUESTION 3.–

The equilibrium equation provides:

$$\begin{cases} \dfrac{\partial\sigma_x}{\partial x}+\dfrac{\partial\tau_{xy}}{\partial y}=0 \\ \dfrac{\partial\tau_{xy}}{\partial x}+\dfrac{\partial\sigma_y}{\partial y}=0 \end{cases} \qquad [10.254]$$

So, assuming the form of σ_x proposed:

$$\begin{cases} \tau_{xy} = K.\dfrac{y^2}{2} + f(x) \\ \sigma_y = f'(x).y + g(x) \end{cases} \quad [10.255]$$

where f and g are two functions of x to be determined.

We then write the boundary conditions on the top and bottom sides:

$$\begin{cases} \tau_{xy}(x, \pm h/2) = 0 \\ \sigma_y(x, \pm h/2) = 0 \end{cases} \quad [10.256]$$

Hence:

$$\begin{cases} f(x) = K.\dfrac{h^2}{8} \\ g(x) = 0 \end{cases} \quad [10.257]$$

And:

$$\begin{cases} \tau_{xy} = \dfrac{K}{2}.\left(y^2 - \dfrac{h^2}{4}\right) \\ \sigma_{yy} = 0 \end{cases} \quad [10.258]$$

And, writing the boundary condition on the right side of the beam:

$$\iint_{SL} \underline{\sigma}(M(L,y), \underline{x}).dS = \int_{y=-h/2}^{h/2} \int_{z=-b/2}^{b/2} \frac{K}{2}.\left(y^2 - \frac{h^2}{4}\right).\underline{y}.dy.dz = F.\underline{y} \quad [10.259]$$

We then obtain the value of K according to F:

$$K = \frac{-12.F}{b.h^3} \quad [10.260]$$

And we can show that the second boundary condition on this side is verified automatically:

$$\iint_{SL} \underline{GM} \wedge \underline{\sigma}\left(M(L,y),\underline{x}\right).dS = \int_{y=-h/2}^{h/2} \int_{z=-b/2}^{b/2} \frac{K}{2}.\left(y^2 - \frac{h^2}{4}\right)\underline{x} \wedge y.\underline{z}.dy.dz = \underline{0} \qquad [10.261]$$

QUESTION 4.–

The form of σ_x is linear function of y, so we have compression (as $K<<0$) in the top section and tension in the lower section; since the linear form is clearly the simplest function to verify this. In addition, σ_x is linear function of x, as bending moment. And σ_x is zero at right side of the beam and maximum in $x = 0$, as bending moment.

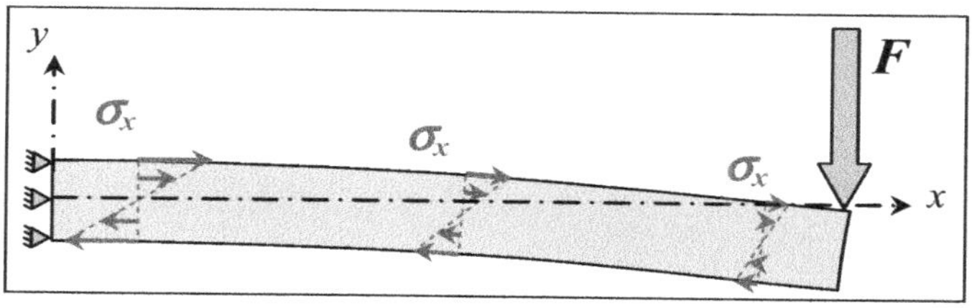

Figure 10.51. *Stress in a bending beam. For a color version of this figure, see www.iste.co.uk/bouvet/aeronautical.zip*

QUESTION 5.–

We plot σ_x and τ_{xy} for any cross-section of x-coordinate:

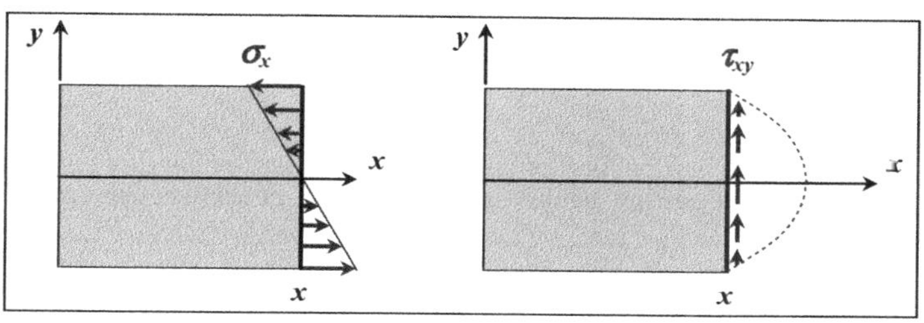

Figure 10.52. *Stress in a bending beam. For a color version of this figure, see www.iste.co.uk/bouvet/aeronautical.zip*

We see σ_x linear function of y, maximum below and minimum on top and a parabolic τ_{xy} zero on top and below, and maximum at the center. The σ_x stress is clearly in the x-direction (facet of normal vector $\underline{x}$) and the τ_{xy} stress is clearly in the y-direction (facet of normal vector $\underline{x}$), even if their parabolic distribution is difficult to make clear, hence the dotted curve drawn in x (however, this does not mean that this stress is in the x-direction!).

Then we calculate the resultant of the stress:

$$\iint_{Sx} \underline{\sigma}(M,\underline{x}).dS = \int_{y=-h/2}^{h/2} \int_{z=-b/2}^{b/2} \left(K.(L-x).\underline{x} + \frac{K}{2}.\left(y^2 - \frac{h^2}{4} \right).\underline{y} \right).dy.dz = F.\underline{y} \quad [10.262]$$

And the moment:

$$\iint_{Sx} \underline{GM} \wedge \underline{\sigma}(M,\underline{x}).dS = \int_{y=-h/2}^{h/2} \int_{z=-b/2}^{b/2} \left(K.(L-x).\underline{x} + \frac{K}{2}.\left(y^2 - \frac{h^2}{4} \right).\underline{y} \right) \wedge y.\underline{y}.dy.dz = F.(L-x)\underline{z} \quad [10.263]$$

Of course, this calculation is much like that performed for the boundary condition in $x = L$.

Clearly, we have found the torsor of internal forces determined in question 1.

QUESTION 6.–

Aluminum is a ductile material. Accordingly, we must choose the Von Mises criterion (or the Tresca criterion):

$$\overline{\sigma} < \sigma_e \quad [10.264]$$

Now we must calculate this Von Mises stress and, crucially, determine where it is maximum:

$$\overline{\sigma} = \sqrt{\frac{3}{2}.\underline{dev}(\underline{\underline{\sigma}}):\underline{dev}(\underline{\underline{\sigma}})} = \sqrt{\sigma_x^2 + 3.\tau_{xy}^2} = |K|.\sqrt{(L-x)^2.y^2 + \frac{3}{4}.\left(y^2 - \frac{h^2}{4} \right)^2} \quad 10.265]$$

Note that here we are in 2D and with $\sigma_y = 0$, hence the simplified expression for Von Mises stress, but this is clearly not generally true.

Under the root we have the total of two positive terms, so this total is maximum at $x = L$ and is:

$$\overline{\sigma}_{\max} = |K|.\sqrt{\frac{3}{4}.y^4 + \left(L^2 - \frac{3.h^2}{8}\right).y^2 + \frac{3.h^4}{64}} \qquad [10.266]$$

Under the root we have the total of three positive terms (as $L > h$), so this total is maximum at $y = \pm h/2$ and is:

$$\overline{\sigma}_{\max} = \frac{|K|.L.h}{2} \qquad [10.267]$$

Essentially, the maximum Von Mises criterion is reached at two points with coordinate $(0, \pm h/2)$ and is of the value above.

QUESTION 7.–

So at the criterion limit, we obtain:

$$|K_{\lim}| = \frac{2.\sigma_e}{K.L.h} \qquad [10.268]$$

Hence:

$$F_{\lim} = \frac{b.h^2.\sigma_e}{6.L} \qquad [10.269]$$

Once this criterion is reached, the beam will begin to develop plasticity at the two points $(0, \pm h/2)$. If we exceed this value, then we will obtain plastic strain and an irreversible deformation of the beam. Indeed, the Von Mises criterion is a criterion for the start of plasticity rather than a failure criterion!

QUESTION 8.–

Glass is a brittle material, so we must choose the criterion of maximum normal stress:

$$Max(\sigma_I, \sigma_{II}, \sigma_{III}) < \sigma_r \qquad [10.270]$$

There remains but to calculate the maximum principal stress and above all where it is maximum:

$$\begin{cases} \sigma_I = \dfrac{\sigma_x}{2} + \sqrt{\left(\dfrac{\sigma_x}{2}\right)^2 + \tau_{xy}{}^2} = \dfrac{K.(L-x).y}{2} + \dfrac{|K|}{2}\sqrt{(L-x)^2.y^2 + \left(y^2 - \dfrac{h^2}{4}\right)^2} \\ \sigma_{II} = \dfrac{\sigma_x}{2} - \sqrt{\left(\dfrac{\sigma_x}{2}\right)^2 + \tau_{xy}{}^2} = \dfrac{K.(L-x).y}{2} - \dfrac{|K|}{2}\sqrt{(L-x)^2.y^2 + \left(y^2 - \dfrac{h^2}{4}\right)^2} \\ \sigma_{III} = 0 \end{cases} \quad [10.271]$$

Note that here we are in 2D and with $\sigma_y = 0$, hence the simplified expression for principal stresses, but this is clearly generally not true. We can also verify in passing that if $\sigma_x = 0$ we indeed find the two principal stresses $\pm\tau_{xy}$ and if $\tau_{xy} = 0$ that we indeed find the two principal stresses σ_x and 0.

Under the root we have the total of two positive terms, so this total is maximum at $x = L$, and, considering that y can have the same positive and negative values, the maximum is obtained for $y < 0$ (since $K < 0$):

$$Max(\sigma_I, \sigma_{II}, \sigma_{III}) = \frac{|K|}{2}\left(-L.y + \sqrt{L^2.y^2 + \left(y^2 - \frac{h^2}{4}\right)^2}\right) \quad \text{with } y < 0 \quad [10.272]$$

Hence:

$$Max(\sigma_I, \sigma_{II}, \sigma_{III}) = \frac{|K|}{2}\left(-L.y + \sqrt{y^4 + \left(L^2 - \frac{h^2}{2}\right).y^2 + \frac{h^4}{16}}\right) \quad \text{with } y < 0 \quad [10.273]$$

Under the root we have the total of three positive terms (since $L > h$), so this total is maximum at $y = -h/2$ (since $y < 0$) and is:

$$\overline{\sigma}_{\max} = \frac{|K|.L.h}{2} \quad [10.274]$$

Essentially, the criterion of maximum normal stress is reached at point $(0, -h/2)$ and is of the value above.

QUESTION 9.–

So, at the criterion limit, we obtain:

$$\left|K_{\lim}\right| = \frac{2.\sigma_r}{K.L.h} \quad [10.275]$$

Hence:

$$F_{\lim} = \frac{b.h^2.\sigma_r}{6.L} \quad [10.276]$$

Once this criterion is reached, the beam will begin to break at point *(0, −h/2)*. Indeed, the criterion of maximum normal stress is a failure criterion! In addition, at the point where $\tau_{xy} = 0$, we will see a crack propagate perpendicular to σ_x and in the *y*-direction.

Part 2: strain tensor and displacement

QUESTION 10.–

Assuming the material is linear elastic, homogeneous and isotropic, we have:

$$\underline{\underline{\varepsilon}} = \frac{1+\nu}{E}\underline{\underline{\sigma}} - \frac{\nu}{E}.trace\left(\underline{\underline{\sigma}}\right).\underline{\underline{I}} \quad [10.277]$$

Hence:

$$\begin{cases} \varepsilon_x = \dfrac{K}{e}.(L-x).y \\ \varepsilon_y = \dfrac{-\nu.K}{e}.(L-x).y \\ \gamma_{xy} = \dfrac{K.(1+\nu)}{E}.\left(y^2 - \dfrac{h^2}{4}\right) \end{cases} \quad [10.278]$$

QUESTION 11.–

If $\nu = 0$:

$$\begin{cases} \varepsilon_x = \dfrac{K}{e}.(L-x).y \\ \varepsilon_y = 0 \\ \gamma_{xy} = \dfrac{K}{E}.\left(y^2 - \dfrac{h^2}{4}\right) \end{cases} \quad [10.279]$$

Then we must integrate the relationship linking strain to displacement:

$$\underline{\underline{\varepsilon}} = \frac{1}{2}.\left(\underline{\underline{grad}}(\underline{u}) + \underline{\underline{grad}}^t(\underline{u})\right) \quad [10.280]$$

Hence:

$$\begin{cases} \dfrac{\partial u}{\partial x} = \dfrac{K}{e}.(L-x).y \\ \dfrac{\partial v}{\partial y} = 0 \\ \dfrac{\partial u}{\partial y} + \dfrac{\partial v}{\partial x} = \dfrac{K}{E}.\left(y^2 - \dfrac{h^2}{4}\right) \end{cases}$$

$$\Rightarrow \begin{cases} u = \dfrac{-K}{2.E}.x^2.y + \dfrac{K}{E}.L.x.y + f(y) \\ v = g(x) \\ \dfrac{-K}{2.E}.x^2 + \dfrac{K}{E}.L.x + f'(y) + g'(x) = \dfrac{K}{E}.\left(y^2 - \dfrac{h^2}{4}\right) \end{cases} \quad [10.281]$$

where $f(y)$ and $g(x)$ are the two functions to determine. The last equation is the sum of terms function only of x and of other terms function only of y, hence:

$$g'(x) - \frac{K}{2.E}.x^2 + \frac{K}{E}.L.x = \frac{K}{E}.\left(y^2 - \frac{h^2}{4}\right) - f'(y) = ct_1 \quad [10.282]$$

where ct_1 is a constant, hence:

$$\begin{cases} u = \dfrac{-K}{2.E}.x^2.y + \dfrac{K}{E}.L.x.y + \dfrac{K}{3.E}.y^3 - \dfrac{K}{E}.\dfrac{h^2}{4}.y - ct_1.y + ct_2 \\ v = \dfrac{K}{6.E}.x^3 - \dfrac{K}{2.E}.L.x^2 + ct_1.x + ct_3 \end{cases} \quad [10.283]$$

And by writing these three boundary conditions, we have:

$$\begin{cases} u = \dfrac{-K}{2.E}.x^2.y + \dfrac{K}{E}.L.x.y + \dfrac{K}{3.E}.y^3 - \dfrac{K.h^2}{12.E}.y \\ v = \dfrac{K}{6.E}.x^3 - \dfrac{K}{2.E}.L.x^2 - \dfrac{K.h^2}{6.E}.x \end{cases} \quad [10.284]$$

QUESTION 12.–

Effectively, we see that the data of these three boundary conditions lets us determine all of the integration constants ct_1, ct_2 and ct_3. Of course, it is not by chance, boundary conditions must prevent all rigid body displacement fields that are in 2D, two translations (in the x and y-directions) and one rotation (in the z-direction). And we observe that the first two boundary conditions prevent translations by clamping the point *(0, 0)* and the third preventing rotation around this point.

QUESTION 13.–

The deflection is defined as the displacement in the y-direction of point *(L, 0)*, so that:

$$\delta = v(L,0) = \frac{-K.L^3}{3.E} - \frac{K.L.h^2}{6.E} \quad [10.285]$$

Hence, if $L >> h$:

$$\delta = \frac{-K.L^3}{3.E} = \frac{4.F.L^3}{E.b.h^3} = \frac{F.L^3}{3.E.I_z} \quad [10.286]$$

NOTE: we obtain $K = -1.92\ N/m^4$

Part 3: finite elements comparison

NOTE ON UNITS.– In a calculations code by finite elements, any unit can be chosen but, we must simply use coherent units. We typically use mm and N, so the unit for stress is MPa.

QUESTION 14.–

Concerning displacements:

If $L << h$ (here we have $L/h = 20$), we obtain:

$$\begin{cases} u = \dfrac{-K}{2.E}.x^2.y + \dfrac{K}{E}.L.x.y \\ v = \dfrac{K}{6.E}.x^3 - \dfrac{K}{2.E}.L.x^2 \end{cases} \qquad [10.287]$$

We do have zero v at clamping point and maximum at $x = L$. The numerical application gives a deflection of 3.2 mm. And we verify that v is not function of y.

We have u that is much less than v (at the first order according to the previous relationship), zero at the clamping point and maximum at $x = L$. And we verify that it is linear function of y and that its maximum value is reached at $(L, h/2)$ and is 0.12 mm, and its minimum value is reached at $(L, -h/2)$ and is −0.12 mm.

Concerning σ_x:

We have zero σ_x at $x = L$ and at $y = 0$, and linear function of y and of x. The numerical application gives a maximum value of 480 MPa at (0, −h/2) and minimum of −480 MPa at $(0, h/2)$. We obtain 485 MPa numerically, but you must not forget that the calculated stress only applies far from the boundary conditions (Saint Venant's principle), since the maximum value is indeed reached at a boundary condition! Unfortunately, this is often the case!

Concerning σ_y:

According to our calculations, we have zero σ_y everywhere! But again, our calculations are only applicable far from the boundary conditions (Saint Venant's principle) where this stress is effectively low. In addition, in the calculation by finite elements, displacements have been imposed as zero where the material is clamped, which is not verified by our analytical model where we impose:

$$\begin{cases} u\,(x=0, y=0)=0 \\ v\,(x=0, y=0)=0 \\ u\,(x=0, y=h/2)=0 \end{cases} \qquad [10.288]$$

We can show that by rigorously imposing zero displacements at the clamping point creates σ_y stress, but the analytical calculation is then much more complex.

Concerning τ_{xy}:

We can verify that τ_{xy} is maximum at $y = 0$, zero at $y = \pm h/2$ and not function of x, of course with the exception at the boundary conditions (Saint Venant's principle). The numerical application provides a maximum value of 6.0 MPa, which is almost verified, given that the scale is adapted to disturbances met at the boundary conditions.

Concerning $\sigma_{VonMises}$:

We obtain a Von Mises stress almost equal to the absolute value of σ_x (since τ_{xy} is negligible). We can analytically obtain a maximum value of 480 MPa. This value is reached at $(0, \pm h/2)$, as obtained analytically.

In practice, we can show that to size a bending beam we must size it to σ_x. This only applies if it is indeed a beam, i.e. if $L >> h$ and if the cross-section is full. Indeed, for hollow beams, tubes for instance, shear stress cannot be disregarded.

Concerning $Max(\sigma_I, \sigma_{II})$:

We obtain a maximum principal stress almost equal to the positive value of σ_x (since τ_{xy} is negligible). We can analytically obtain a maximum value of 480 MPa. This value is reached at $(0, \pm h/2)$, as obtained analytically.

Concerning $Min(\sigma_I, \sigma_{II})$:

We obtain a minimum principal stress almost equal to the negative value of σ_x (since τ_{xy} is negligible). We can analytically obtain a minimum value of −480 MPa. This value is reached at $(0, h/2)$.

Appendix

Analysis Formulas

A.1. Analysis formulas in Cartesian coordinates

Gradient of a scalar field:

$$\underline{grad}(f) = \frac{\partial f}{\partial x}.\underline{x} + \frac{\partial f}{\partial y}.\underline{y} + \frac{\partial f}{\partial z}.\underline{z} = f_{,i} = \underline{\nabla}.f \quad \text{[A.1]}$$

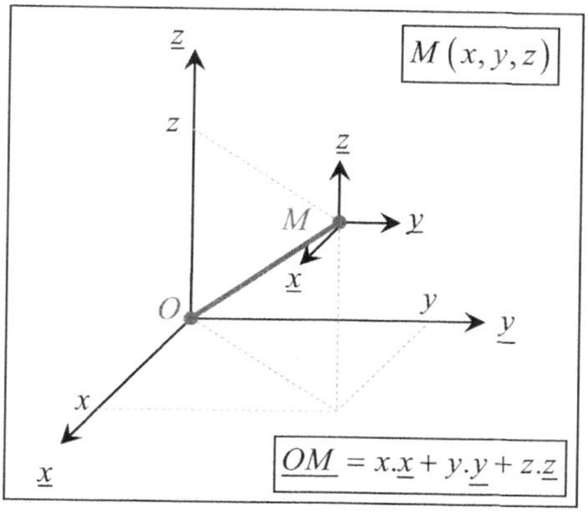

Figure A.1. *Coordinates of a point in Cartesian coordinates. For a color version of this figure, see www.iste.co.uk/bouvet/aeronautical.zip*

Divergence of a vector field:

$$div(\underline{V}) = \frac{\partial V_x}{\partial x} + \frac{\partial V_y}{\partial y} + \frac{\partial V_z}{\partial z} = trace\left(\underline{grad}(f)\right) = f_{i,i} = \underline{\nabla}.\underline{V} \quad \text{[A.2]}$$

Laplacian of a scalar field:

$$\Delta f = \frac{\partial^2 f}{\partial x^2} + \frac{\partial^2 f}{\partial y^2} + \frac{\partial^2 f}{\partial z^2} = f_{,ii} \qquad [A.3]$$

Rotational of a vector field :

$$\underline{\underline{rot}}\,(\underline{V}) = \left[\frac{\partial V_z}{\partial y} - \frac{\partial V_y}{\partial z}\right]\underline{x} + \left[\frac{\partial V_x}{\partial z} - \frac{\partial V_z}{\partial x}\right]\underline{y} + \left[\frac{\partial V_y}{\partial x} - \frac{\partial V_x}{\partial y}\right]\underline{z} = \underline{\nabla} \wedge \underline{V} \qquad [A.4]$$

Gradient tensor:

$$\underline{\underline{grad}}\,(\underline{V}) = \begin{bmatrix} \frac{\partial V_x}{\partial x} & \frac{\partial V_x}{\partial y} & \frac{\partial V_x}{\partial z} \\ \frac{\partial V_y}{\partial x} & \frac{\partial V_y}{\partial y} & \frac{\partial V_y}{\partial z} \\ \frac{\partial V_z}{\partial x} & \frac{\partial V_z}{\partial y} & \frac{\partial V_z}{\partial z} \end{bmatrix} = V_{i,j} = \underline{\underline{\nabla}}\,\underline{V} \qquad [A.5]$$

Divergence of a tensor field:

$$\underline{div}\left(\underline{\underline{T}}\right) = \begin{bmatrix} \frac{\partial T_{xx}}{\partial x} + \frac{\partial T_{xy}}{\partial y} + \frac{\partial T_{xz}}{\partial z} \\ \frac{\partial T_{yx}}{\partial x} + \frac{\partial T_{yy}}{\partial y} + \frac{\partial T_{yz}}{\partial z} \\ \frac{\partial T_{zx}}{\partial x} + \frac{\partial T_{zy}}{\partial y} + \frac{\partial T_{zz}}{\partial z} \end{bmatrix} = T_{ij,j} = \underline{\nabla}.\underline{\underline{T}} \qquad [A.6]$$

Also note that (and this is true irrespective of the coordinate type):

$$\underline{\Delta}\,\underline{u} = \underline{\underline{grad}}(div\,\underline{u}) - \underline{rot}(\underline{rot}\,\underline{u}) \qquad [A.7]$$

Laplacian of a vector field:

$$\underline{\Delta}\,\underline{V} = \begin{bmatrix} \Delta V_x \\ \Delta V_y \\ \Delta V_z \end{bmatrix} \qquad [A.8]$$

Laplacian of a tensor field:

$$\underline{\underline{\Delta\sigma}} = \begin{bmatrix} \Delta\sigma_x & sym & sym \\ \Delta\tau_{xy} & \Delta\sigma_y & sym \\ \Delta\tau_{xz} & \Delta\tau_{yz} & \Delta\sigma_z \end{bmatrix} \quad [A.9]$$

A.2. Analysis formulas in cylindrical coordinates

Gradient of a scalar field:

$$\underline{grad}(f) = \frac{\partial f}{\partial r}.\underline{r} + \frac{1}{r}\cdot\frac{\partial f}{\partial \theta}.\underline{\theta} + \frac{\partial f}{\partial z}.\underline{z} \quad [A.10]$$

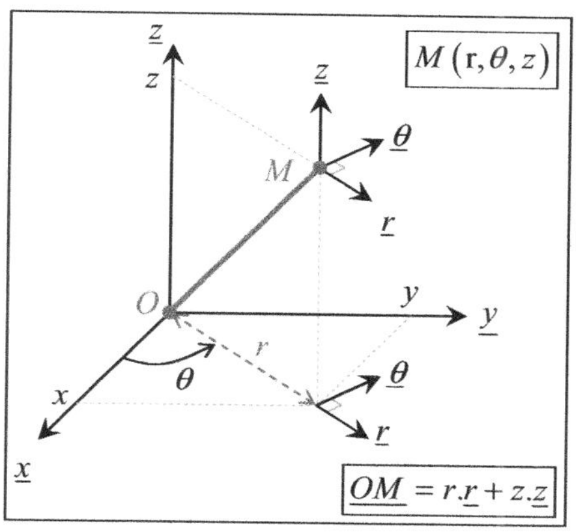

Figure A.2. *Coordinates of a point in cylindrical coordinates. For a color version of this figure, see www.iste.co.uk/bouvet/aeronautical.zip*

Divergence of a vector field:

$$div(\underline{V}) = \frac{V_r}{r} + \frac{\partial V_r}{\partial r} + \frac{1}{r}\cdot\frac{\partial V_\theta}{\partial \theta} + \frac{\partial V_z}{\partial z} \quad [A.11]$$

Laplacian of a scalar field:

$$\Delta f = \frac{\partial^2 f}{\partial r^2} + \frac{1}{r}\cdot\frac{\partial f}{\partial r} + \frac{1}{r^2}\cdot\frac{\partial^2 f}{\partial \theta^2} + \frac{\partial^2 f}{\partial z^2} \quad [A.12]$$

Rotational of a vector field:

$$\underline{rot}\ \underline{V} = \left[\frac{1}{r}\cdot\frac{\partial V_z}{\partial \theta} - \frac{\partial V_\theta}{\partial z}\right]\underline{r} + \left[\cdot\frac{\partial V_r}{\partial z} - \frac{\partial V_z}{\partial r}\right]\underline{\theta} + \left[\frac{V_\theta}{r} + \frac{\partial V_\theta}{\partial r} - \frac{1}{r}\cdot\frac{\partial V_r}{\partial \theta}\right]\underline{z} \qquad [A.13]$$

Gradient tensor:

$$\underline{\underline{grad}}\left(\underline{V}\right) = \begin{bmatrix} \frac{\partial V_r}{\partial r} & \frac{1}{r}\cdot\frac{\partial V_r}{\partial \theta} - \frac{V_\theta}{r} & \frac{\partial V_r}{\partial z} \\ \frac{\partial V_\theta}{\partial r} & \frac{1}{r}\cdot\frac{\partial V_\theta}{\partial \theta} + \frac{V_r}{r} & \frac{\partial V_\theta}{\partial z} \\ \frac{\partial V_z}{\partial r} & \frac{1}{r}\cdot\frac{\partial V_z}{\partial \theta} & \frac{\partial V_z}{\partial z} \end{bmatrix} \qquad [A.14]$$

Divergence of a tensor field:

$$\underline{div}\left(\underline{\underline{T}}\right) = \begin{bmatrix} \frac{\partial T_{rr}}{\partial r} + \frac{1}{r}\cdot\frac{\partial T_{\theta r}}{\partial \theta} + \frac{\partial T_{zr}}{\partial z} + \frac{T_{rr} - T_{\theta\theta}}{r} \\ \frac{\partial T_{r\theta}}{\partial r} + \frac{1}{r}\cdot\frac{\partial T_{\theta\theta}}{\partial \theta} + \frac{\partial T_{z\theta}}{\partial z} + \frac{T_{r\theta} + T_{\theta r}}{r} \\ \frac{\partial T_{rz}}{\partial r} + \frac{1}{r}\cdot\frac{\partial T_{\theta z}}{\partial \theta} + \frac{\partial T_{zz}}{\partial z} + \frac{T_{rz}}{r} \end{bmatrix} \qquad [A.15]$$

Laplacian of a vector field:

$$\underline{\Delta}\ \underline{V} = \begin{bmatrix} \Delta V_r - \frac{V_r}{r^2} - \frac{2}{r^2}\frac{\partial V_\theta}{\partial \theta} \\ \Delta V_\theta + \frac{2}{r^2}\frac{\partial V_r}{\partial \theta} - \frac{V_\theta}{r^2} \\ \Delta V_z \end{bmatrix} \qquad [A.16]$$

Laplacian of a tensor field:

$$\underline{\underline{\Delta}}\ \underline{\underline{\sigma}} = \begin{bmatrix} \Delta\sigma_r - \frac{2}{r^2}\left(2\frac{\partial \tau_{r\theta}}{\partial \theta} + \sigma_r - \sigma_\theta\right) & sym & sym \\ \Delta\tau_{r\theta} + \frac{2}{r^2}\left(\frac{\partial(\sigma_r - \sigma_\theta)}{\partial \theta} - 2\tau_{r\theta}\right) & \Delta\sigma_\theta + \frac{2}{r^2}\left(2\frac{\partial \tau_{r\theta}}{\partial \theta} + \sigma_r - \sigma_\theta\right) & sym \\ \Delta\tau_{rz} - \frac{1}{r^2}\left(2\frac{\partial \tau_{\theta z}}{\partial \theta} + \tau_{rz}\right) & \Delta\tau_{\theta z} + \frac{1}{r^2}\left(2\frac{\partial \tau_{rz}}{\partial \theta} + \tau_{\theta z}\right) & \Delta\sigma_z \end{bmatrix} \qquad [A.17]$$

A.3. Analysis formulas in spherical coordinates

Gradient of a scalar field:

$$\underline{grad}(f)=\frac{\partial f}{\partial \rho}.\underline{\rho}+\frac{1}{\rho}\cdot\frac{\partial f}{\partial \varphi}.\underline{\varphi}+\frac{1}{\rho.\sin\varphi}\cdot\frac{\partial f}{\partial \theta}.\underline{\theta} \quad [A.18]$$

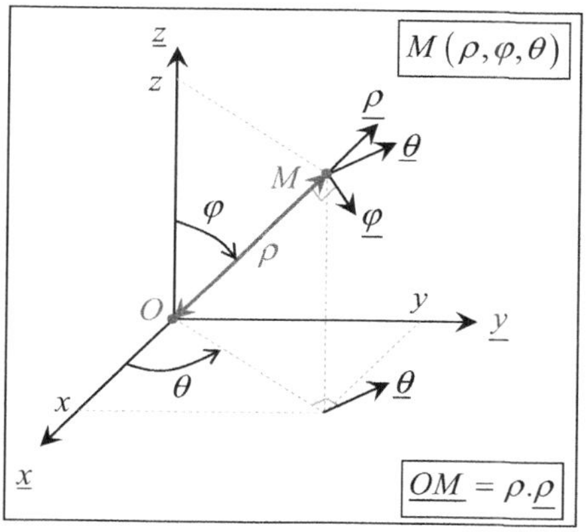

Figure A.3. *Coordinates of a point in spherical coordinates. For a color version of this figure, see www.iste.co.uk/bouvet/aeronautical.zip*

Divergence of a vector field:

$$div(\underline{V})=\frac{1}{\rho^2}\cdot\frac{\partial}{\partial \rho}\left(\rho^2.V_\rho\right)+\frac{1}{\rho.\sin\varphi}\cdot\frac{\partial}{\partial \varphi}\left(V_\varphi.\sin\varphi\right)+\frac{1}{\rho.\sin\varphi}\cdot\frac{\partial V_\theta}{\partial \theta} \quad [A.19]$$

Laplacian of a scalar field:

$$\Delta f=\frac{\partial^2 f}{\partial \rho^2}+\frac{2}{\rho}\cdot\frac{\partial f}{\partial \rho}+\frac{1}{\rho^2}\cdot\frac{\partial^2 f}{\partial \varphi^2}+\frac{1}{\rho^2.\tan\varphi}\cdot\frac{\partial f}{\partial \varphi}+\frac{1}{\rho^2.\sin^2\varphi}\cdot\frac{\partial^2 f}{\partial \theta^2} \quad [A.20]$$

Rotational of a vector field:

$$\begin{aligned}\underline{\text{rot}}\ \underline{\text{V}}=&\frac{1}{\rho^2.\sin\varphi}\cdot\left[\frac{\partial}{\partial \varphi}(\rho.V_\theta.\sin\varphi)-\frac{\partial}{\partial \theta}\left(\rho.V_\varphi\right)\right].\underline{\rho}\\&+\frac{1}{\rho.\sin\varphi}\cdot\left[\frac{\partial V_\rho}{\partial \theta}-\frac{\partial}{\partial \rho}(\rho.V_\theta.\sin\varphi)\right].\underline{\varphi}+\frac{1}{\rho}\cdot\left[\frac{\partial}{\partial \rho}\left(\rho.V_\varphi\right)-\frac{\partial V_\rho}{\partial \varphi}\right].\underline{\theta}\end{aligned} \quad [A.21]$$

Gradient tensor:

$$\underline{\underline{grad}}\left(\underline{V}\right)=\begin{bmatrix} \dfrac{\partial V_\rho}{\partial\rho} & \dfrac{\partial V_\varphi}{\partial\rho} & \dfrac{\partial V_\theta}{\partial\rho} \\ \dfrac{1}{\rho}\cdot\dfrac{\partial V_\rho}{\partial\varphi}-\dfrac{V_\varphi}{\rho} & \dfrac{1}{\rho}\cdot\dfrac{\partial V_\varphi}{\partial\varphi}+\dfrac{V_\rho}{\rho} & \dfrac{1}{\rho}\cdot\dfrac{\partial V_\theta}{\partial\varphi} \\ \dfrac{1}{\rho.\sin\varphi}\cdot\dfrac{\partial V_\rho}{\partial\theta}-\dfrac{V_\theta}{\rho} & \dfrac{1}{\rho.\sin\varphi}\cdot\dfrac{\partial V_\varphi}{\partial\theta}-\dfrac{V_\theta}{\rho.\tan\varphi} & \dfrac{1}{\rho.\sin\varphi}\cdot\dfrac{\partial V_\theta}{\partial\theta}+\dfrac{V_\rho}{\rho}+\dfrac{V_\varphi}{\rho.\tan\varphi} \end{bmatrix} \quad [A.22]$$

Divergence of a tensor field:

$$\underline{div}\left(\underline{\underline{T}}\right)=\begin{bmatrix} \dfrac{\partial T_{\rho\rho}}{\partial\rho}+\dfrac{1}{\rho}\cdot\dfrac{\partial T_{\varphi\rho}}{\partial\varphi}+\dfrac{1}{\rho.\sin\varphi}\cdot\dfrac{\partial T_{\theta\rho}}{\partial\theta}+\dfrac{2.T_{\rho\rho}-T_{\theta\theta}-T_{\varphi\varphi}+T_{\varphi\rho}.\cot\varphi}{\rho} \\ \dfrac{\partial T_{\rho\varphi}}{\partial\rho}+\dfrac{1}{\rho}\cdot\dfrac{\partial T_{\varphi\varphi}}{\partial\varphi}+\dfrac{1}{\rho.\sin\varphi}\cdot\dfrac{\partial T_{\theta\varphi}}{\partial\theta}+\dfrac{2.T_{\rho\varphi}+T_{\varphi\rho}+\left(T_{\varphi\varphi}-T_{\theta\theta}\right).\cot\varphi}{\rho} \\ \dfrac{\partial T_{\rho\theta}}{\partial\rho}+\dfrac{1}{\rho}\cdot\dfrac{\partial T_{\varphi\theta}}{\partial\varphi}+\dfrac{1}{\rho.\sin\varphi}\cdot\dfrac{\partial T_{\theta\theta}}{\partial\theta}+\dfrac{2T_{\rho\theta}+T_{\theta\rho}+\left(T_{\varphi\theta}+T_{\theta\varphi}\right).\cot\varphi}{\rho} \end{bmatrix} \quad [A.23]$$

Laplacian of a vector field:

$$\underline{\Delta V}=\begin{bmatrix} \Delta V_\rho-2.\dfrac{V_\rho}{\rho^2}-\dfrac{2}{\rho^2}\cdot\dfrac{\partial V_\varphi}{\partial\varphi}-\dfrac{2.V_\varphi}{\rho^2.\tan\varphi}-\dfrac{2}{\rho^2.\sin\varphi}\cdot\dfrac{\partial V_\theta}{\partial\theta} \\ \Delta V_\varphi+\dfrac{2}{\rho^2}\cdot\dfrac{\partial V_\rho}{\partial\varphi}-\dfrac{V_\varphi}{\rho^2.\sin^2\varphi}-\dfrac{2.\cos\varphi}{\rho^2.\sin^2\varphi}\cdot\dfrac{\partial V_\theta}{\partial\theta} \\ \Delta V_\theta+\dfrac{2}{\rho^2.\sin\varphi}\cdot\dfrac{\partial V_\rho}{\partial\theta}+\dfrac{2.\cos\varphi}{\rho^2.\sin^2\varphi}\cdot\dfrac{\partial V_\varphi}{\partial\theta}-\dfrac{V_\theta}{\rho^2.\sin^2\varphi} \end{bmatrix} \quad [A.24]$$

with:

$$\Delta=\frac{\partial^2}{\partial\rho^2}+\frac{2}{\rho}\cdot\frac{\partial}{\partial\rho}+\frac{1}{\rho^2}\cdot\frac{\partial^2}{\partial\varphi^2}+\frac{1}{\rho^2.\tan\varphi}\cdot\frac{\partial}{\partial\varphi}+\frac{1}{\rho^2.\sin^2\varphi}\cdot\frac{\partial^2}{\partial\theta^2} \quad [A.25]$$

Bibliography

[AGA 08] AGATI P., LEROUGE F., ROSSETTO M., *Résistance des matériaux, Cours, exercices et applications industrielles*, Dunod, 2008.

[ASH 80] ASHBY M.F., JONES D.R.H., *Engineering Materials 1: An Introduction to Properties, Applications and Design*, Elsevier, 1980.

[ASH 86] ASHBY M.F., JONES D.R.H., *Engineering Materials 2: An introduction to microstructures and processing*, Elsevier, 1986.

[BAM 08] BAMBEGER Y., VOLDOIRE F., *Mécanique des structures: Initiations, approfondissements, applications*, Presses des Ponts et Chaussées, 2008.

[BER 99] BERTHELOT J.M., *Matériaux composites*, Tec & Doc, 1999.

[BOU 01] BOUVET C., De l'Uniaxial au Multiaxial: Comportement Pseudo-élastique des Alliages à Mémoire de Forme, PhD Thesis, University of Besançon, 2001.

[BOU 16] BOUVET C., Tolérance aux dommages d'impact des structures composites aéronautiques, article TRP4042, *Techniques de l'ingénieur*, 2016.

[BRO 04] BRON F., Déchirure ductile des tôles minces en alliage d'aluminium 2024 pour application aéronautique, PhD Thesis, Mines de Paris, 2004.

[CAS 13] CASTANIE B., BOUVET C., GUEDRA-DEGEORGES D., Structures en matériaux composites stratifiés, *Techniques de l'ingénieur*, 2013.

[CHE 08] CHEVALIER L., *Mécanique des systèmes et des milieux déformables: Cours, exercices et problèmes corrigés*, Ellipses, 2008.

[COI 01] COIRIER J., *Mécanique des milieux continus, cours et exercices corrigés*, Dunod, 2001.

[DEL 08] DELAPLACE A., GATUINGT F., RAGUENEAU F., *Aide-mémoire: Mécanique des structures*, Dunod, 2008.

[DEQ 12] DEQUATREMARE M., DEVERS T., *Précis des matériaux*, Dunod, 2012.

[DEV 86] DEVELAY R., Traitements thermiques des alliages d'aluminium, *Techniques de l'ingénieur*, 1986.

[DOR 86] DORLOT J.M., BAÏLON J.P., MASOUNAVE J., *Des matériaux*, Edition de l'école Polytechnique de Montréal, 1986.

[DUP 09] DUPEUX M., *Introduction à la mécanique des matériaux et des structures*, Dunod, 2009.

[DUP 13] DUPEUX M., *Science des matériaux*, Dunod, 2013.

[DUV 98] DUVAUT G., *Mécanique des milieux continus*, Dunod, 1998.

[FRA 12] FRANÇOIS D., PINEAU A., ZAOUI A., *Mechanical Behaviour of Materials*, Springer Netherlands, 2012.

[GAY 15] GAY D., *Matériaux composites*, 6th ed. Hermès-Lavoisier, 2015.

[GER 73] GERMAIN P., *Cours de mécanique des milieux continus, tome 1: théorie générale*, Masson, 1973.

[GOU 10] GOURGUES-LORENZON A.F., HAUDIN J.M., *Matériaux pour l'ingénieur*, Presses des Mines de Paris, 2010.

[HEA 77] HEARN E.J., *Mechanics of Materials*, Pergamon International Library, 1977.

[KHA 95] KHAN A.S., HUANG S., *Continuum Theory of Plasticity*, John Wiley & Sons, 1995.

[LEM 96] LEMAITRE J., CHABOCHE J.L., *Mechanics of Solid Materials*, Cambridge University Press, 1996.

[MIR 03] MIRANVILLE A., TEMAM R., *Modélisation mathématique et mécanique des milieux continus*, Springer-Verlag, 2003.

[SAL 01] SALENÇON J., *Handbook of Continuum Mechanics*, Springer-Verlag, Berlin Heidelberg, 2001.

[THU 97] THUAL O., *Introduction à la mécanique des milieux continus déformables*, Cépaduès, 1997.

[UGU 03] UGURAL A.C., FENSTER S.K., *Advanced Mechanics of Materials and Applied Elasticity*, Prentice Hall, 2003.

Index

D, E, F

G, H, I

J, L, M

N, O, P

Q, R, S

T, U

V, W, Y

Other titles from

ISTE

in

Mechanical Engineering and Solid Mechanics

2017

BOREL Michel, VÉNIZÉLOS Georges
Movement Equations 2: Mathematical and Methodological Supplements (Non-deformable Solid Mechanics Set – Volume 2)

EL HAMI Abdelkhalak, RADI Bouchaib
Fluid-Structure Interactions and Uncertainties: Ansys and Fluent Tools

KHARMANDA Ghias, EL HAMI Abdelkhalak
Biomechanics: Optimization, Uncertainties and Reliability (Reliability of Multiphysical Systems Set – Volume 5)

LEDOUX Michel, EL HAMI Abdelkhalak
Fluid Mechanics: Analytical Methods(Mathematical and Mechanical Engineering Set – Volume 3)

Compressible Flow Propulsion and Digital Approaches in Fluid Mechanics (Mathematical and Mechanical Engineering Set – Volume 4)

MORI Yvon
Mechanical Vibrations: Applications to Equipment

2016

BOREL Michel, VÉNIZÉLOS Georges
Movement Equations 1: Location, Kinematics and Kinetics (Non-deformable Solid Mechanics Set – Volume 1)

BOYARD Nicolas
Heat Transfer in Polymer Composite Materials

CARDON Alain, ITMI Mhamed
New Autonomous Systems (Reliability of Multiphysical Systems Set – Volume 1)

DAHOO Pierre Richard, POUGNET Philippe, EL HAMI Abdelkhalak
Nanometer-scale Defect Detection Using Polarized Light (Reliability of Multiphysical Systems Set – Volume 2)

DE SAXCÉ Géry, VALLÉE Claude
Galilean Mechanics and Thermodynamics of Continua

DORMIEUX Luc, KONDO Djimédo
Micromechanics of Fracture and Damage (Micromechanics Set – Volume 1)

EL HAMI Abdelkhalak, RADI Bouchaib
Stochastic Dynamics of Structures (Mathematical and Mechanical Engineering Set – Volume 2)

GOURIVEAU Rafael, MEDJAHER Kamal, ZERHOUNI Noureddine
From Prognostics and Health Systems Management to Predictive Maintenance 1: Monitoring and Prognostics (Reliability of Multiphysical Systems Set – Volume 4)

KHARMANDA Ghias, EL HAMI Abdelkhalak
Reliability in Biomechanics (Reliability of Multiphysical Systems Set – Volume 3)

MOLIMARD Jérôme
Experimental Mechanics of Solids and Structures

RADI Bouchaib, EL HAMI Abdelkhalak
Material Forming Processes: Simulation, Drawing, Hydroforming and Additive Manufacturing (Mathematical and Mechanical Engineering Set – Volume 1)

2015

KARLIČIĆ Danilo, MURMU Tony, ADHIKARI Sondipon, MCCARTHY Michael
Non-local Structural Mechanics

SAB Karam, LEBÉE Arthur
Homogenization of Heterogeneous Thin and Thick Plates

2014

ATANACKOVIC M. Teodor, PILIPOVIC Stevan, STANKOVIC Bogoljub, ZORICA Dusan
Fractional Calculus with Applications in Mechanics: Vibrations and Diffusion Processes

ATANACKOVIC M. Teodor, PILIPOVIC Stevan, STANKOVIC Bogoljub, ZORICA Dusan
Fractional Calculus with Applications in Mechanics: Wave Propagation, Impact and Variational Principles

CIBLAC Thierry, MOREL Jean-Claude
Sustainable Masonry: Stability and Behavior of Structures

ILANKO Sinniah, MONTERRUBIO Luis E., MOCHIDA Yusuke
The Rayleigh–Ritz Method for Structural Analysis

LALANNE Christian
Mechanical Vibration and Shock Analysis – 5-volume series – 3rd edition
Sinusoidal Vibration – volume 1
Mechanical Shock – volume 2
Random Vibration – volume 3
Fatigue Damage – volume 4
Specification Development – volume 5

LEMAIRE Maurice
Uncertainty and Mechanics

2013

ADHIKARI Sondipon
Structural Dynamic Analysis with Generalized Damping Models: Analysis

ADHIKARI Sondipon
Structural Dynamic Analysis with Generalized Damping Models: Identification

BAILLY Patrice
Materials and Structures under Shock and Impact

BASTIEN Jérôme, BERNARDIN Frédéric, LAMARQUE Claude-Henri
Non-smooth Deterministic or Stochastic Discrete Dynamical Systems: Applications to Models with Friction or Impact

EL HAMI Abdelkhalak, BOUCHAIB Radi
Uncertainty and Optimization in Structural Mechanics

KIRILLOV Oleg N., PELINOVSKY Dmitry E.
Nonlinear Physical Systems: Spectral Analysis, Stability and Bifurcations

LUONGO Angelo, ZULLI Daniele
Mathematical Models of Beams and Cables

SALENÇON Jean
Yield Design

2012

DAVIM J. Paulo
Mechanical Engineering Education

DUPEUX Michel, BRACCINI Muriel
Mechanics of Solid Interfaces

ELISHAKOFF Isaac *et al.*
Carbon Nanotubes and Nanosensors: Vibration, Buckling and Ballistic Impact

GRÉDIAC Michel, HILD François
Full-Field Measurements and Identification in Solid Mechanics

GROUS Ammar
Fracture Mechanics – 3-volume series
Analysis of Reliability and Quality Control – volume 1
Applied Reliability – volume 2
Applied Quality Control – volume 3

RECHO Naman
Fracture Mechanics and Crack Growth

2011

KRYSINSKI Tomasz, MALBURET François
Mechanical Instability

SOUSTELLE Michel
An Introduction to Chemical Kinetics

2010

BREITKOPF Piotr, FILOMENO COELHO Rajan
Multidisciplinary Design Optimization in Computational Mechanics

DAVIM J. Paulo
Biotribolgy

PAULTRE Patrick
Dynamics of Structures

SOUSTELLE Michel
Handbook of Heterogenous Kinetics

2009

BERLIOZ Alain, TROMPETTE Philippe
Solid Mechanics using the Finite Element Method

LEMAIRE Maurice
Structural Reliability

2007

GIRARD Alain, ROY Nicolas
Structural Dynamics in Industry

GUINEBRETIÈRE René
X-ray Diffraction by Polycrystalline Materials

KRYSINSKI Tomasz, MALBURET François
Mechanical Vibrations

KUNDU Tribikram
Advanced Ultrasonic Methods for Material and Structure Inspection

SIH George C. *et al.*
Particle and Continuum Aspects of Mesomechanics

Printed and bound by CPI Group (UK) Ltd, Croydon, CR0 4YY

13/09/2026

14961039-0005